Studies in Systems, Decision and Control

Volume 312

Series Editor

Janusz Kacprzyk, Systems Research Institute, Polish Academy of Sciences, Warsaw, Poland

The series "Studies in Systems, Decision and Control" (SSDC) covers both new developments and advances, as well as the state of the art, in the various areas of broadly perceived systems, decision making and control–quickly, up to date and with a high quality. The intent is to cover the theory, applications, and perspectives on the state of the art and future developments relevant to systems, decision making, control, complex processes and related areas, as embedded in the fields of engineering, computer science, physics, economics, social and life sciences, as well as the paradigms and methodologies behind them. The series contains monographs, textbooks, lecture notes and edited volumes in systems, decision making and control spanning the areas of Cyber-Physical Systems, Autonomous Systems, Sensor Networks, Control Systems, Energy Systems, Automotive Systems, Biological Systems, Vehicular Networking and Connected Vehicles, Aerospace Systems, Automation, Manufacturing, Smart Grids, Nonlinear Systems, Power Systems, Robotics, Social Systems, Economic Systems and other. Of particular value to both the contributors and the readership are the short publication timeframe and the world-wide distribution and exposure which enable both a wide and rapid dissemination of research output.

** Indexing: The books of this series are submitted to ISI, SCOPUS, DBLP, Ulrichs, MathSciNet, Current Mathematical Publications, Mathematical Reviews, Zentralblatt Math: MetaPress and Springerlink.

More information about this series at http://www.springer.com/series/13304

Paweł J. Mitkowski

Mathematical Structures of Ergodicity and Chaos in Population Dynamics

 Springer

Paweł J. Mitkowski
Kraków, Poland

ISSN 2198-4182   ISSN 2198-4190   (electronic)
Studies in Systems, Decision and Control
ISBN 978-3-030-57680-6   ISBN 978-3-030-57678-3   (eBook)
https://doi.org/10.1007/978-3-030-57678-3

This Springer imprint is published by the registered company Springer Nature Switzerland AG
The registered company address is: Gewerbestrasse 11, 6330 Cham, Switzerland

I believe that mathematics is simply the structure of the world. Not a description of this structure, but the structure itself.

— Prof. ANDRZEJ LASOTA

taken from (A. Lasota and A. Klimek 2005)

To my Parents and Kasia

Preface

The presented monograph is based on the doctoral dissertation of the author (Mitkowski 2011), (see also Mitkowski 2012) in the field of biocybernetics and biomedical engineering presented in 2011 at AGH—University of Science Technology in Kraków, Poland. This field includes issues related to medicine, biology, biomathematics or cybernetics as practiced by engineers. Considered there population dynamics models are still in the interest of researchers, and even this interest is increasing, especially now in the time of SARS-CoV-2 coronavirus pandemic, when models are intensively studied in order to help predict its behaviour within human population.

The presented monograph is written from the perspective of technical sciences, engineering and applications. That is, basing on mathematical foundations, the author presents population dynamics models. Identifies practically chaotic dynamics in the strictly defined sense in spaces important from applied sciences point of view. The identification was made using a computational, engineering approach to show symptoms of chaotic behaviour experimentally. The results were interpreted on the background of real medical data. In general, the research was developed with an idea to create the possibility of comparing it with the measurements of real phenomena. However, the issues studied here have a broader meaning than the area of application. That is why the preliminary cognitive remarks were also presented. The keynote of the book is words of the mathematician Prof. Andrzej Lasota: "I believe that mathematics is simply the structure of our world. Not a description of this structure, but the structure itself." (Lasota and Klimek 2005).

In Chap. 1, the philosophical basis of research presented in the monograph is presented. That is, two frequently occurring approaches to understanding what mathematics is, and hence what mathematical modeling is discussed. The contraction mapping theorem of Stefan Banach, which defines the fundamental principles of scientific thinking is also stated with proof that is carried out using elementary methods. Then, in Chap. 2, biological and medical data on the red blood cell (erythrocyte) dynamics are discussed. Particularly describing individual stages of the erythropoiesis process, i.e. the production of erythrocytes. Further in Chap. 3, a set of selected mathematical concepts is provided in order to direct the reader to important groups of issues from the point of view of the presented research. In Chap. 4, ideas of researching chaos on the basis of ergodic theory is presented. Then in Sect. 4.2 a systematic discussion of mathematical theory concerning chaos in systems that maintain measure is shown. In addition, in Chap. 4 calculation examples showing the methodology of numerical studies on low-dimensional mapping are included. In Chap. 6 they are appropriately transferred to the infinite-dimensional system in a form of delay differential equation. In Chap. 5, the theory of the classic Lasota-Ważewska equation is presented, which is necessary to understand the principles of the infinite-dimensional model mentioned above. Chapter 6 includes computational verification of Lasota hypothesis of nontrivial ergodic properties existence (what means chaotic behaviour) in his delay differential equation with unimodal feedback.

During the work on the research presented in this monograph, many people provided me substantive consultations and help. I express my sincere gratefulness to my Father Prof. Wojciech Mitkowski from the Department of Automatics of AGH—University of Science and Technology (AGH-UST), Cracow, Poland. I would like to thank Prof. Maciej J. Ogorzałek from Jagiellonian University (JU), Cracow, Poland, who was the promoter of my Ph.D. thesis, Prof. Ryszard Rudnicki from Institute of Mathematics of Polish Academy of Sciences (IMPAS), Warsaw, Poland, Prof. Antoni Leon Dawidowicz from JU, Prof. Józef Myjak from AGH-UST, Prof. Michał Rams from IMPAS, Prof. Zbigniew Galias from AGH-UST, Prof. Piotr Rusek from AGH-UST, Prof. Aleksander Skotnicki Head of the Hematology Clinic of Collegium Medicum of the JU in the years 1993–2018, Prof. Andrzej Szczeklik former Head of the Second Department and Clinic of Internal Medicine of Collegium Medicum of the JU and former Vice President of the Polish Academy of Art and Sciences (PAAS) in Kraków, Prof. Andrzej Białas former President of the PAAS in Kraków, Poland, Ewa Gudowska-Nowak from the Department of Physics of the JU, Dr. Zofia Mitkowska from the University Children's Hospital, Cracow, Poland, Dr. Teresa Wolska-Smoleń from the Hematology Clinic of Collegium Medicum of the JU, Dr. Barbara Sokołowska from the Second Department and Clinic of Internal Medicine of Collegium Medicum of the JU and Dr. Robert Szczelina for Department of Mathematics of JU. I would like to thank Prof. Eduardo Liz from the Department of Applied Mathematics II of the University of Vigo, Spain for the discussion regarding delay differential equations, what in some sense became the inspiration to write this book.

I express my deep gratitude to the Series Editor Prof. Janusz Kacprzyk from the Systems Research Institute, Polish Academy of Sciences, Warsaw, Poland for the key support of the this book publication. Further more I wish to thank Dr. Thomas Ditziger, Springer Editorial Director, Dieter Merkle, Guido Zosimo-Landolfo, Sylvia Schneider, Prasanna Kumar. N and Monica Janet, M. from Springer Nature for kind guiding and help in the publishing process.

Nothing could be done without the support of my Parents and Kasia.

Warsaw, Poland   Paweł J. Mitkowski
May 2020

Contents

Chapter 1
Introduction

I will begin this book by discussing two cognitive issues that seem to be the foundation for my further considerations. First of all, I will discuss two frequently cited definitions of mathematics that are important here. Next, I will present a general philosophy of conducting scientific research, the structure of which can be found in the contraction mapping theorem of Stefan Banach.

1.1 Mathematical Study of the Reality Structure

Many centuries ago it was already noticed that in order to learn the true structure of the world, it is necessary to study it through geometry and numbers (see e.g. Penrose 2010; Tatarkiewicz 1978, volumes 1-3). We intuitively feel that geometric concepts such as figure, volume, surface, distance etc. naturally fit into the analysis of e.g. the spatial structure of reality. It is also necessary to quantify the size of these elements in space, and numbers are used naturally here. In modern mathematics, we have many more and much more subtle structures that seem to be developments of these primal ones, e.g. in various complicated spaces. Over the centuries, mathematical concepts have been more and more precise, and thus whole new branches of mathematics have been created, the source of which, however, has always been, and still is, the quest for better learning about structures of reality.

Mathematics is often defined in two different ways. Some people think that it is the language to describe the world, i.e. it has a similar function as the languages we speak, e.g. Polish, English, Spanish, etc. When we want to describe our observations, we choose appropriate words for this purpose, although we often realize that our description does not reflect the complexity and richness of the phenomenon that we describe. Still, while mathematics is considered to be the language for an approximate description of the world, there is another approach to mathematics, called Platonic. It goes further by defining mathematics as simply the structure of the world, not

P. J. Mitkowski, *Mathematical Structures of Ergodicity and Chaos in Population Dynamics*, Studies in Systems, Decision and Control 312, https://doi.org/10.1007/978-3-030-57678-3_1

just the language to describe it. With this assumption, mathematical research (e.g., mathematical modeling of dynamics) becomes a search for structures existing in the surrounding reality. In contrast to the Platonic approach, in the philosophy of mathematics there is so-called intuitionism. This is intuitionism as defined by the Dutch mathematician Brouwer (see Penrose 1996, p. 134), according to which one cannot speak about the existence of mathematical beings in reality until one gives a way to develop them.

1.2 Foundations of Scientific Thinking

Using mathematical reasoning, any problem we analyze can have one solution, a finite number of solutions (more than one), infinitely many solutions, or may not have a solution at all. It is worth to know before starting, e.g. complicated and time-consuming computer calculations, whether the problem that we are studying computationally has a solution at all, or how many of these solutions we should expect. Reasoning as such seems to be fundamental from the perspective of conducting any scientific research, understood here as finding the truth about the phenomenon or object of reality being examined. The essence of scientific thinking can be found in the contraction mapping theorem of Stefan Banach (Mitkowski 2006, p. 13; Kudrewicz 1976, p. 43; Rudnicki 2001, p. 94), (see e.g. Musielak 1976, p. 196).

1.2.1 Banach Contraction Mapping Theorem

Let (X, ρ) be a metric space with the metric ρ (see Sect. 3.1). A mapping $F: X \rightarrow X$ is called Lipschitz map with the Lipschitz constant $\alpha > 0$, if

$$\forall x_1 \in X \; \forall x_2 \in X \; \rho(F(x_1), F(x_2)) \leq \alpha \rho(x_1, x_2). \qquad (1.1)$$

Condition (1.1) is called Lipschitz condition. A map $F: X \rightarrow X$ is called contraction, if it meets the condition (1.1) with a constant $\alpha < 1$. Point $x^* \in X$ is called a fixed point of the map $F: X \rightarrow X$, if $F(x^*) = x^*$.

Theorem 1.1 (Banach) *Let (X, ρ) be a complete metric space (see Sect. 3.1), and $F: X \rightarrow X$ be the contraction. Then, there is exactly one fixed point x^* of the map F. In addition, if*

$$x_{n+1} = F(x_n), \quad n = 0, 1, 2, \ldots \qquad (1.2)$$

then for each $x_0 \in X$

$$\lim_{n \to \infty} x_n = x^* \qquad (1.3)$$

and

$$\rho(x_m, x^*) \le \frac{\alpha^m}{1 - \alpha} \, \rho(x_0, F(x_0)). \tag{1.4}$$

Proof Let x_0 be any point of a space X and $x_{n+1} = F(x_n)$ for $n = 0, 1, 2, \ldots$. Condition (1.1) results in the following inequalities

$$\rho(x_1, x_2) = \rho(F(x_0), F(x_1)) \le \alpha\rho(x_0, x_1),$$

$$\rho(x_2, x_3) = \rho(F(x_1), F(x_2)) \le \alpha\rho(x_1, x_2) \le \alpha^2\rho(x_0, x_1),$$

$$\vdots$$

$$\rho(x_n, x_{n+1}) \le \alpha^n \rho(x_0, x_1) \quad \text{dla } n = 0, 1, 2, \ldots .$$

Let us now estimate the distance $\rho(x_n, x_m)$ for $n < m$. From the triangle inequality we have

$$\rho(x_n, x_m) \le \rho(x_n, x_{n+1}) + \rho(x_{n+1}, x_{n+2}) + \cdots + \rho(x_{m-1}, x_m) \le$$

$$\le (\alpha^n + \alpha^{n+1} + \cdots + \alpha^{m-1})\rho(x_0, x_1) \le \frac{\alpha^n}{1 - \alpha} \, \rho(x_0, x_1). \tag{1.5}$$

Because $\alpha < 1$, so for any $\epsilon > 0$, there exists $n_0 \in N$ (where N is the set of natural numbers) such that

$$\frac{\alpha^n}{1 - \alpha} \, \rho(x_0, x_1) < \epsilon \quad \text{for } n_0 \le n. \tag{1.6}$$

So (from (1.5) and (1.6)) $\rho(x_n, x_m) < \epsilon$ for $n_0 \le n$, hence the sequence $\{x_n\}$ is the Cauchy sequence (see Sect. 3.1) (one can also write that $\rho(x_n, x_m) \to 0$, at $n, m \to \infty$). Space (X, ρ) is complete, so there exists a limit $x^* = \lim_{n \to \infty} x_n$. We, therefore have

$$\rho(x^*, F(x^*)) \le \rho(x^*, x_n) + \rho(x_n, F(x^*)) \le \rho(x^*, x_n) + \alpha\rho(x_{n-1}, x^*) \to \infty$$

at $n \to \infty$. This means $\rho(x^*, F(x^*)) = 0$, so $x^* = F(x^*)$. In addition, let us suppose that $y^* = F(y^*)$. If $\rho(x^*, y^*) > 0$, we would have

$$\rho(x^*, y^*) = \rho(F(x^*), F(y^*)) \le \alpha\rho(x^*, y^*) < \rho(x^*, y^*),$$

which is impossible. Therefore, $\rho(x^*, y^*) = 0$, this means $x^* = y^*$. This proves the uniqueness of the solution x^* of the equation $x^* = F(x^*)$. Let us check also the inequality (1.4) from the thesis of Theorem 1.1. From (1.5), we receive

$$\rho(x_m, x^*) = \lim_{n \to \infty} \rho(x_m, x_n) \leq \frac{\alpha^m}{1 - \alpha} \rho(x_0, x_1). \tag{1.7}$$

The contraction map theorem determines the condition that must be fulfilled so that the considered problem has exactly one solution. In addition, it provides the algorithm of how to approach this solution "in practice" and specifies the estimation of the distance from the solution at each step of the algorithm. This principle defines the essence of scientific thinking. One can clearly see, how mathematics and algorithmic methods (e.g. applied science methods) differ in terms of cognition. It is solely through mathematics that we can prove the existence of a solution. Well-designed algorithmic methods can only bring us closer to the solution, but it is obvious that in a finite time (after a finite number of iterations) we will never reach the exact solution. It could be said that the exact solution cannot be "materially touched", but only approached. In practice, however, it is enough only to "approach" to the solution. After all, we cannot mathematically define the entire structure of the world around us, and yet we are able to construct many complicated technical devices that are useful. Therefore, it seems that both pathways of cognition, experimental (physical) and structural (mathematical), functioning together are important.

1.3   Mathematical Modelling in Biology and Medicine

Let us now move on to the issues that this monograph directly concerns, i.e. to the mathematical modelling of biological and medical processes. There are several stages of modelling (see Rudnicki 2014, p. 9), (and also Ważewska-Czyżewska 1983, p. 155), (as well as Foryś 2005; Murray 2006):

1. Determining biological and medical assumptions using mathematical concepts.
2. Providing an appropriate mathematical model, e.g. function, differential equation, etc.
3. Examining the model properties using mathematical and/or numerical methods.
4. Biological and medical interpretation of theoretical (mathematical and numerical) results.
5. Comparing theoretical results with real measurements. Providing possible correction of the model in order to obtain better match to real data. Possible proposals of new experimental studies for further model verification.

Specialists in biomathematics quite often emphasize (see e.g. Rudnicki 2014, p. 9), (and also Ważewska-Czyżewska 1983, p. 155), (as well as Foryś 2005, p. 17) that skillful simplification of the model is a key procedure in the modeling process. The essence is not to replace the complicated real biological and medical process with an equally complicated mathematical model, but to identify the most important biological and medical premises and include them in a possibly simple mathematical model (sometimes reduced from one that is more complex), which can then be effectively studied through mathematical methods. Model reduction (see Rudnicki 2014, p. 9) leads to a partial description of the studied phenomenon that often does not entirely reflect its complexity. However, such a reduced model, if well constructed, may show the important structure (e.g. dynamic) inherent in the studied real process, which is responsible for some specific nature of its changes in time.

1.4 Topics and Objectives of the Book

Lasota (1977) noted that the irregular waveforms observed in the measurement results of real biological and medical processes can be associated not only with the high complexity of these processes and the imperfections of measurement methods, but also with the very structure of these systems. In other words, real biological and medical processes may simply be chaotic under certain conditions, i.e. they may "contain" a mathematical structure that generates chaos. Note that depending on the adoption of one of the two approaches to mathematics described in the this chapter, the process of mathematical modeling can be considered as building a description of this structure using mathematical language, or its direct search within the system. Lasota believed that such mathematical structure "responsible" for chaos in biological systems can be invariant measures with non-trivial ergodic properties (see Chap. 4), we could also say that the system exhibits non-trivial ergodic properties. Lasota formalized this approach for his proposed model of changes in the number of circulating red blood cells in the bloodstream, putting forward a hypothesis about the non-trivial ergodic properties of this model (see Chap. 6). The model is in the form of a delay differential equation in which, with appropriate selection of parameters, the function of erythrocyte production can be set in unimodal form (with one smooth maximum—the unimodal function is precisely described in Definition 6.22). Such nature of the dependence of the erythrocyte production intensity on their quantity in the bloodstream medically corresponds to the pathological conditions that occur in the body. It is also possible when the body is in a near-death state (Ważewska-Czyżewska 1983, pp. 165–167). Normal erythropoiesis (production of erythrocytes) is characterized by monotonically decreasing dependence. On this assumption, a lot of mathematical models were constructed and studied, among others, the classic Lasota-Ważewska model, about which we will write extensively in the further part of the book (see Chap. 5). In this work, we are interested in the irregular dynamics of the model with unimodal erythopoiesis in the form that the system has three stationary points (differences between the studied model and other biological models with unimodal nonlinearities are presented in Chap. 6). Specifically, we want to computationally verify the hypothesis of Lasota, which in the language of ergodic theory raises the problem of the existence of chaotic solutions for a certain class of delay differential equations. In addition, it seems to be a generalization (for such class of equations) of the earlier hypothesis proposed by Ulam (1960) about the non-trivial ergodic properties of $[0, 1]$ interval mapping into itself (see Sect. 4.3 and Chap. 6). It is important that this class of equations has very large applications in modeling biological processes. Evidence for the existence of periodic solutions of such equations is known (see Chow 1974; Kaplan and Yorke 1977), (see also Ważewska-Czyżewska 1983; Rudnicki 2014; Alvarez et al. 2019), (in terms of computer-assisted proofs see Szczelina and Zgliczyński 2018; Lessard and Mireles James 2019). Yet until now, apart from special cases (see Walther 1981; Walther 1999; Lani-Wayda 1999), little can be said about the existence of chaotic solutions in a precisely defined sense for delay differential equations in general. We will conduct research on chaotic behavior using numerical methods, therefore, we are aware that the results obtained can only be an indication of some type of chaos and will not provide proof of its existence in a mathematical sense. However, from the point of view of technical sciences, it seems to us that the prepared calculations will be a valuable example of how, using a numerical experiment, one can examine the existence of certain mathematical structures "responsible" for chaos in a specific way. In recent years, computer-assisted proofs are being strongly developed (see e.g. Tucker 1999; Galias and Zgliczński 1998; Galias 2003; Szczelina and Zgliczyński 2018; Lessard and Mireles James 2019). From a bio-medical point of view, obtaining solutions indicating the nontrivial ergodic properties of the model (and thus indicating the existence of chaotic solutions in a certain sense) will suggest that with disturbed erythropoietic response in unimodal form, irregular changes may occur in the quantity of erythrocytes circulating in the bloodstream. Such irregular changes mean a strong disturbance in the functioning of the entire red blood cell system, which, when working properly, has a very strong tendency to maintain erythrocytes quantity at a constant level (Ważewska-Czyżewska and Lasota 1976, p. 24). In addition to the numerical analysis of chaos, we expand the bio-medical interpretation of the studied equation, among others by providing the meaning of one of its parameters s (see Sect. 6.1), which according to our knowledge was given first by Mitkowski (2011), (see also Mitkowski 2012), and on whose value depends whether the erythropoietic response is physiologically correct (monotonically decreasing) or distorted to an unimodal function.

Chapter 2
Dynamics of the Red Blood Cell System

The bio-medical data developed in this chapter were derived primarily from: (Ważewska-Czyżewska 1983; Ważewska-Czyżewska and Lasota 1976; Lasota et al. 1981; Mackey and Milton 1990; Craig et al. 2010; Rudnicki 2009; Keener and Sneyd 1998).

2.1 Blood

Blood consists of two main elements; liquid plasma (constituting about 55% of its total volume) and suspended in plasma cell components (about 40% of volume). Cellular components are divided into three categories: **erythrocytes** (red blood cells), **leukocytes** (white blood cells) and **thrombocytes** (platelets).

2.2 Erythrocytes

Erythrocytes (red blood cells) are shaped like small, double-sided concave discs with a diameter of about 8 μm. They transport oxygen from the lungs to the tissues and carbon dioxide in the opposite direction. Erythrocytes contain a protein called hemoglobin (the red blood pigment) that is responsible for the gas exchange between tissues and lungs. They do not have a cell nucleus or mitochondria; they are flexible and can pass through small capillary vessels. Red blood cells do not divide and do not have a mechanism that could repair the damage that occurs in them over time. In a healthy person, their lifetime is about 120 days, after which they are destroyed mainly in the spleen (less frequently in the liver). The body must, therefore, constantly produce new erythrocytes to replace those that have broken down. Under disease conditions, red blood cells may die after a shorter period of time, what causes so-

P. J. Mitkowski, *Mathematical Structures of Ergodicity and Chaos
in Population Dynamics*, Studies in Systems, Decision and Control 312,
https://doi.org/10.1007/978-3-030-57678-3_2

called anemia. Furthermore, certain diseases in which young erythrocytes or even their precursors (cells from which they are formed) are destroyed can occur, which leads to ineffective erythropoiesis. Erythropoiesis is the process of producing red blood cells (we will refer to this later in the Sect. 2.4). In an adult human, 1 mm^3 of blood contains 4.2–5.4 million erythrocytes, while in a newborn, it is about 7 million.

2.3 Bone Marrow

Bone marrow is a soft, highly blood-filled, spongy tissue located inside the long bone marrow cavities and in small cavities within the cancellous bone. It accounts for about 5% of body weight in a healthy adult human. There are two types of bone marrow; the so-called yellow bone marrow consisting mainly of adipose cells (this type of bone marrow is hematologically inactive) and the so-called red bone marrow, which produces erythrocytes, leukocytes and thrombocytes. In the post-natal child, red bone marrow fills all the bones. With age, it is replaced by yellow bone marrow, and in adults, red bone marrow is present only in the bones of the sternum, spine, pelvis, ribs, clavicles, skull, humerus and femur. The amount of red bone marrow may increase when blood cells are needed.

In the bone marrow, certain **hematopoietic stem cells (HSC)** exist, also called **pluripotential hematopoietic stem cells (PPSC)**. A group of mature cells that are used in case of increased demand for morphotic elements is also located there. Normal bone marrow has a characteristic organization; there are so-called nests of erythrocyte precursors accumulating around a centrally located macrophage that stores iron, there are also large cells called megakaryocytes producing and releasing platelets into the vessels, and near the trabeculae there are cells responsible for production of white blood cells. Mature cells migrate from the marrow cavities towards the vascular sinuses.

Some studies suggest (Craig et al. 2010, p. 929) that the bone marrow contains stem cells that can transform into other than hematopoietic cells, e.g. nerve, muscle, skeletal and cardiomyocytes (myocardial cells), as well as hepatic and vascular endothelial cells.

2.4 Erythropoiesis—The Production of Erythrocytes

In the body of a healthy human being, staying in stable conditions (Ważewska-Czyżewska and Lasota 1976, p. 24), there is a strong tendency to maintain number of circulating erythrocytes at a constant level. Reduction of their amount triggers the process of forming erythrocytes in the bone marrow called **erythropoiesis**. Generally, a change in the number of red blood cells in the bloodstream causes stimulation or inhibition of their production (Ważewska-Czyżewska and Lasota 1976, p. 26). Response of the system (e.g. in the form of an increase in the number of erythrocytes after their previous decrease) appears after the time needed to transform the undifferentiated bone marrow cell into a mature erythrocyte. Therefore, the red blood cell production system can be interpreted as a delayed argument feedback system. The time needed for the production of mature erythrocytes according to Ważewska-Czyżewska and Lasota (1976, p. 23) and Ważewska-Czyżewska (1983, p. 162) is 3–4 days. Mackey and Milton (1990, p. 6) and Mackey (1997, p. 156) state an average of 5.7 days, whereas Keener and Sneyd (1998, p. 491) a time of 5–7 days, but not less than 5 days, even at a high stimulation rate of red blood cell production. These differences may result from the fact that the starting moment during erythropoiesis from which the time starts to count is determined differently (see detailed description of erythropoiesis stages below). E.g. Lasota et al. (1981, p. 151) hold that a signal stimulating the production of erythropoietin, as a result of a decrease in oxygen concentration in the arteries, is this starting moment.

Erythropoiesis in an adult healthy person occurs in the bone marrow. In the case of the fetus, it proceeds in other locations at different stages of its growth. In adults with significant disorders of the structure and functioning of the bone marrow, it can occur in the liver or in the spleen.

2.4.1   Stages of Erythrocyte Development

2.4.1.1   Stem Cells

The youngest erythrocyte precursor cell is an **undifferentiated hematopoietic bone marrow stem cell** (pluripotent stem cell—**PPSC**). This cell has the ability to self-renew and to pass into the production lines of erythrocytes, leukocytes and thrombocytes. At steady state, most PPSC cells are in the G_0 resting phase or in the G_1 active phase of the cell cycle. From phase G_0, PPSC can shift into the cell division cycle, or they can differentiate into **precursors of one of the blood cell differentiation and maturation lines**. We are going to discuss the red blood cell line.

2.4.1.2   Progenitor Cells

Differentiation of erythrocyte line precursors takes place in several stages (Ważewska-Czyżewska 1983, p. 25); **the primitive BFU-E cell** (primitive burst-forming unit-erythroid) is described as the least differentiated. Subsequent stages of development include the **mature BFU-E cell** (mature burst-forming unit-erythroid), the **transitional cell** and the **CFU-E cell** (colony-forming unit-erythroid). Differentiation of these cells is induced by the hormone **erythropoietin (Epo)**, which regulates the erythropoiesis process.

The resulting population of **differentiated erythrocyte precursor cells** is uniquely oriented to the formation of cells that, by losing the ability to proliferate (divide and thus renew their pool), gain the ability to carry oxygen. Two different processes occur simultaneously in this population; maturation and proliferation. The maturation process is responsible for creating a functionally efficient "final" cell in the production line, i.e. in the considered case of mature erythrocytes, while the proliferation process ensures pool of diverse erythrocyte precursors replenishment. Maturation and proliferation processes are interrelated, and when the cell reaches a certain maturity level, proliferation stops.

It is also worth noting the differences between the concepts of cell differentiation and maturation. Cell differentiation is an irreversible process of coding new information into a developing cell, while cell maturity can be identified with its level of morphological development.

2.4.1.3 Precursor Cells

The next stages in the development of maturing differentiated erythrocyte precursors are **proerythroblast, basophilic erythroblast, polychromatic erythroblast, orthochromatic erythroblast** and **reticulocyte**. The most important feature in the maturation of these cells is the synthesis of hemoglobin, which beginning means that the differentiation process has been achieved. Disturbances in the synthesis of hemoglobin can cause ineffective erythropoiesis. The transition between polychromatic and orthochromatic erythroblast marks the point where proliferation property is lost (Lasota et al. 1981, p. 150).

2.4.1.4 Peripheral Blood Cells

The cell nucleus is removed from the erythroblast, after which the cell remains as a **reticulocyte** in the bone marrow. The maturation time of bone marrow reticulocyte is about 50 h, the Epo hormone may cause reduction of this time (Ważewska-Czyżewska 1983, p. 58). Reticulocyte leave the bone marrow and, as mature erythrocyte, enter the bloodstream through the walls of the myeloid sinuses.

2.4.1.5 Erythropoietin (Epo)

As we have already mentioned **Erythropoietin (Epo)** is a hormone that regulates erythropoiesis. Its most important function is to induce differentiation of precursors of the erythrocyte production line into a series of erythroblastic cells. In an adult human, 80–90% of Epo is produced in the kidneys, the rest is produced, among others, in the liver. In response, for example, to a decrease in oxygen concentration in the arteries, as a result of a decrease in the number of erythrocytes circulating in the bloodstream, the kidneys synthesize and release erythropoietin.

2.4.2 Mechanisms for Erythropoiesis Regulation

The influx of cells from the PPSC population to the CFU-E population, giving rise to the erythrocyte production line, is controlled by two types of mechanism; the so-called short-range mechanism and the long-range mechanism.

2.4.2.1 Short-Range Mechanism

Short-range mechanisms relate to dynamics within the population of PPSC and CFU-E cells, e.g. a decrease in the number of PPSC cells or an increased level of their differentiation into one of the blood cell production lines results in an increase in the level of proliferation (multiplication) of the PPSC population, while an increase in the number of PPSC cells causes inhibition of their proliferation. Thus, short-range mechanisms are responsible, e.g. for maintaining steady state in the bone marrow stem cell population. Disorders of these mechanisms cause a hematological disease called periodic hematopoiesis (PH) (see Mackey and Milton 1990, p. 17; Ważewska-Czyżewska and Lasota 1976; Ważewska-Czyżewska 1983; Rudnicki and Wieczorek 2009), which is characterized by the occurrence of 17–28-day periodic oscillations in the amount of formed blood elements. Hematopoiesis is the process of forming erythrocytes, leukocytes and thrombocytes, therefore, erythropoiesis is one of its subprocesses.

2.4.2.2 Long-Range Mechanism

When the oxygen concentration in the arteries is insufficient, the kidneys synthesize and release erythropoietin (see Ważewska-Czyżewska 1983, p. 139; Lasota et al. 1981, p. 151). This not only induces differentiation of erythrocyte precursors to erythroblasts, but also affects the kinetics of all cell groups in the erythrocyte production system. As a result, the level of red blood cell production increases, among other things, by increasing the maturation rate of their precursors. Therefore, the signal stimulating the operation of the long-range erythropoietic feedback loop is the discrepancy between the tissue's need for oxygen and its supply by erythrocytes circulating in the bloodstream. The existence of feedback between the bone marrow and tissues was first suggested by Paul Bert in 1878 (see Ważewska-Czyżewska 1983, p. 139). He stated that survival at high altitudes requires increased production of erythrocytes to supply the tissues with the right amount of oxygen. Long-range erythropoietic feedback is implemented with the help of the central nervous system (see Ważewska-Czyżewska 1983, p. 139).

Chapter 3
Mathematical Basics

We will now present selected mathematical concepts relevant to the issues discussed in this book.

Section 3.1 was developed based on (Rudnicki 2001; Myjak 2006; Mitkowski 2006; Kudrewicz 1976; Bronsztejn et al. 2004, p. 659). Whereas Sects. 3.2 and 3.3 were developed based on (Lasota and Mackey 1994; Rudnicki 2001; Myjak 2006; Bronsztejn et al. 2004).

3.1 Metric Spaces

Definition 3.1 Non-empty set X with a map $\rho \colon X \times X \to [0, \infty)$ fulfilling the conditions

1. $\rho(x, y) = 0 \Leftrightarrow x = y$;
2. $\rho(x, y) = \rho(y, x)$ for $x, y \in X$;
3. $\rho(x, z) \leq \rho(x, y) + \rho(y, z)$ for $x, y, z \in X$

is called the metric space. The elements of the metric space are called points, the map ρ is the metric in X, and the map value $\rho(x, y)$ is the distance of the x and y points in the metric ρ.

Example 3.1 Let X be any non-empty set. Function

$$\rho(x, y) = \begin{cases} 1 & \text{for} \quad x \neq y \\ 0 & \text{for} \quad x = y \end{cases}$$

is a metric in X, called a discrete metric. Space (X, ρ) is called a discrete space.

P. J. Mitkowski, *Mathematical Structures of Ergodicity and Chaos in Population Dynamics*, Studies in Systems, Decision and Control 312, https://doi.org/10.1007/978-3-030-57678-3_3

Example 3.2 Function

$$\rho(x, y) = |y - x| \quad \text{dla} \quad x, y \in R$$

is a metric in the set of real numbers R.

Example 3.3 In space R^n, $x = (x_1, \ldots, x_n)$, $y = (y_1, \ldots, y_n)$ functions:

$$\rho_1(x, y) = \sqrt{\sum_{k=1}^{n}(x_k - y_k)^2};$$

$$\rho_2(x, y) = \sum_{k=1}^{n}|x_k - y_k|;$$

$$\rho_3(x, y) = \max_{1 \le k \le n}|x_k - y_k|;$$

are metrices. The $\rho_1(x, y)$ metrice is called Euclidean metric, and the space (R^n, ρ_1) is called Euclidean space. In R^2, the above metrics will have the form:

$$\rho_1(x, y) = \sqrt{(x_1 - y_1)^2 + (x_2 - y_2)^2};$$

$$\rho_2(x, y) = |x_1 - y_1| + |x_2 - y_2|;$$

$$\rho_3(x, y) = \max\{|x_1 - y_1|, |x_2 - y_2|\}.$$

Example 3.4 A set of all functions $x(t)$, $y(t)$ continuous in the interval $[a, b]$, with a metric

$$\rho(x, y) = \max_{a \le t \le b}|x(t) - y(t)|$$

is called a space $C(a, b)$.

Definition 3.2 Let (X, ρ) be a metric space. We say that a sequence of points $\{x_n\}$ of that space fulfills the Cauchy condition (or is a Cauchy sequence), if for any ϵ there exists a natural number n_0 such that $\rho(x_m, x_n) < \epsilon$ for $m \ge n_0$ and $n \ge n_0$.

Definition 3.3 A metric space (X, ρ) is called complete, if each sequence satisfying the Cauchy condition is convergent in this space.

Definition 3.4 A set D is dense in X, if $\bar{D} = X$, where $\bar{D}$ is the closure of set D.

3.2 Measures and Measure Spaces

The family of sets will be called a set, the elements of which include certain sets. Let X be any non-empty set.

Definition 3.5 A family $\mathcal{A}$ of sub-sets of set X is called the algebra of sets if:

1. $\emptyset \in \mathcal{A}$;
2. When $A \in \mathcal{A}$ then $X \setminus A \in \mathcal{A}$;
3. When $A \in \mathcal{A}$ and $B \in \mathcal{A}$, then $A \cup B \in \mathcal{A}$.

Definition 3.6 An algebra of sets $\mathcal{A}$ is called the σ-algebra of sets, if for any sets $A_n \in \mathcal{A}$, $n = 1, 2, 3, \ldots$, we have

$$\bigcup_{n=1}^{\infty} A_n \in \mathcal{A}. \tag{3.1}$$

Therefore, σ-algebra $\mathcal{A}$ is a family of subsets of set X satisfying the Conditions 1 and 2 from the Definition 3.5 and the condition (3.1) from the Definition 3.6. Condition 3 from the Definition 3.5 is obtained from the Condition 1 from this definition and from the condition (3.1) from the Definition 3.6. Elements of σ-algebra $\mathcal{A}$ are called measurable sets, because a measure is defined for them.

Definition 3.7 The real-valued function μ defined on σ-algebra $\mathcal{A}$ is a measure if:

1. $\mu(\emptyset) = 0$;
2. $\mu(A) \geq 0$ for all $A \in \mathcal{A}$;
3. $\mu(\cup_k A_k) = \sum_k \mu(A_k)$ if $\{A_k\}$ is finite or infinite sequence of pair-wise disjoint sets from $\mathcal{A}$, that is, $A_i \cap A_j = \emptyset$ for $i \neq j$.

We do not exclude the possibility that $\mu(A) = \infty$ for some $A \in \mathcal{A}$.

Definition 3.8 If $\mathcal{A}$ is a σ-algebra of subsets of a set X and if μ is a measure on $\mathcal{A}$, then $(X, \mathcal{A}, \mu)$ is called measure space.

Remark 3.1 Let Γ be any non-empty family of subsets of a set X. Then there exists the smallest σ-algebra containing Γ.

Remark 3.2 The smallest σ-algebra $\mathcal{A}$ containing a non-empty family Γ of subsets of set X is called σ-algebra generated by the family Γ.

Remark 3.3 Let (X, ρ) be a metric space, and Γ be a family of open (or closed) sets. Then σ-algebra generated by the family Γ is called σ-algebra of the Borel sets $\mathcal{B}$, and its elements are called Borel sets. Definition of σ-algebra of Borel sets depends on the choice of metrice ρ, because metric determines what open sets are.

σ-algebra of Borel sets is the smallest σ-algebra containing all open, thus also closed subsets X.

Remark 3.4 If $X = [0, 1]$ or R (set of real numbers), then the most natural σ-algebra is σ-algebra of Borel sets $\mathcal{B}$. There exists, on σ-algebra of Borel sets, a unique measure μ called the Borel measure, such that $\mu([a, b]) = b - a$.

In applications, the following measure space is used:

Definition 3.9 The measure space $(X, \mathcal{A}, \mu)$ is σ-finite if there exists a sequence $\{A_k\}$, $A_k \in \mathcal{A}$, satisfying

$$X = \bigcup_{k=1}^{\infty} A_k \quad \text{and} \quad \mu(A) < \infty \qquad \text{for all } k.$$

Remark 3.5 If $X = R$ (set of real numbers) and μ is Borel measure, then A_k can be specified in the form of intervals $[-k, k]$, and in d-dimensional space R^d in the form of spheres with radius k.

Definition 3.10 The measure space $(X, \mathcal{A}, \mu)$ is finite if $\mu(X) < \infty$. In particular, if $\mu(X) = 1$, then measure space is normalized or probabilistic.

Definition 3.11 We say, that a certain property occurs almost everywhere (we will write a.e.) if a set of these x, for which it is not satisfied has a measure 0.

3.3   Lebesgue Measure and Integral

We will give the theorem by which we will define the Lebesgue measure (see Rudnicki 2001, p. 458).

Theorem 3.2 *Let $X = R^n$ and let $\mathcal{A}$ be σ-algebra of Borel sets in X. Then there exists exactly one measure λ_n specified on $\mathcal{A}$ such that for any rectangle*

$$P = [a_1, b_1] \times \cdots \times [a_n, b_n]$$

we have

$$\lambda_n(P) = (b_1 - a_1) \cdot \ldots \cdot (b_n - a_n).$$

The measure defined in the Theorem 3.2 is called Lebesgue measure on Borel sets. The Lebesgue measure is a generalization of the concept of volume on Borel sets.

Definition 3.12 Let X be a non-empty set, $\mathcal{A}$ be a σ-algebra on X and $\bar{R} = R \cup \{-\infty, \infty\}$. Function $f : X \to \bar{R}$ is called measurable, if a set

$$\{x \in X : f(x) > a\}$$

is measurable with any $a \in R$.

Theorem 3.3 *If X is a metric space, and $\mathcal{A}$ is a σ-algebra of Borel sets, then any continuous function $f: X \to R$ is measurable.*

Let X be a fixed set, and $\mathcal{A}$ σ-algebra of subsets of set X. For any set $A \subset X$ function

$$1_A = \begin{cases} 1, & \text{when } x \in A, \\ 0, & \text{when } x \notin A \end{cases} \tag{3.2}$$

is called characteristic function of set A. If $A \in \mathcal{A}$, then the characteristic function 1_A is measurable.

Let $A_1, \ldots, A_n$ be any subsets of X, and $c_1, \ldots, c_n$ be any real numbers. Function of the form

$$f(x) = \sum_{i=1}^{n} c_i 1_{A_i}(x)$$

is called simple function. If sets $A_1, \ldots, A_n$ are measurable, then the simple function is measurable.

Definition 3.13 Let $(X, \mathcal{A}, \mu)$ be a measure space. Let $A_1, \ldots, A_n, E$ be measurable sets, and $c_1, c_2, \ldots, c_n$ positive real numbers. Lebesgue integral from a simple function

$$f(x) = \sum_{i=1}^{n} c_i 1_{A_i}(x)$$

over a set E with respect to a measure μ is called a number

$$\sum_{i=1}^{n} c_i \mu(A_i \cap E) \tag{3.3}$$

and is denoted

$$\int_E f(x) d\mu(dx). \tag{3.4}$$

If the function f is measurable and non-negative, and E is a measurable set, then we assume

$$\int_E f d\mu(dx) = \sup \left\{ \int_E g d\mu(dx) : 0 \le g \le f, \ g - \text{simple function} \right\}. \tag{3.5}$$

Let

$$f^+(x) = \max(0, f(x)) \quad \text{i} \quad f^-(x) = \max(0, -f(x)).$$

Note that

$$f(x) = f^+(x) - f^-(x).$$

If at least one of integrals $\int_E f^+ d\mu(dx), \int_E f^- d\mu(dx)$ is finite, then the expression

$$\int_E f d\mu(dx) = \int_E f^+ d\mu(dx) - \int_E f^- d\mu(dx) \qquad (3.6)$$

is called Lebesgue integral from a function f over a set E. We say then, that the function f has the Lebesgue integral on the set E. If the Lebesgue integral is finite, then we say that the function f is integrable in the Lebesgue sense on the set E. If f is integrable on X, then we simply say that f is integrable in the sense of Lebesgue.

Definition 3.14 Let $(X, \mathcal{A}, \mu)$ be a measure space and let p be a real number $1 \leq p \leq \infty$. Family of all possible measurable real functions $f : X \to R$ satisfying

$$\int_X |f(x)|^p \mu(dx) < \infty \qquad (3.7)$$

is a space $L^p(X, \mathcal{A}, \mu)$ (we will write L^p briefly).

When $p = 1$, then a space L^1 consists of all possible integrable functions.
   The norm f in a space L^p is defined by the following expression

$$||f||_{L^p} = \left[\int_X |f(x)|^p \mu(dx) \right]^{1/p}. \qquad (3.8)$$

Chapter 4
Chaos and Ergodic Theory

In the literature concerning dynamic systems, there are many definitions of chaos and several different approaches to its study (see e.g. Rudnicki 2004; Devaney 1987; Bronsztejn et al. 2004). In this book, we will be interested particularly in the approach where chaos is studied using ergodic theory tools (Lasota 1979; Rudnicki 1985a, 2009, 2004; Dawidowicz 1992a). The moment when L. Boltzmann formulated the ergodic hypothesis can probably be considered to be the beginning of ergodic theory. It was in 1868 (see e.g. Nadzieja 1996; Górnicki 2001) or in 1871 according to Lebowitz and Penrose (1973).

4.1 The Gibbs Sets and the Simulations of States Density Evolution

Ergodic theory has its roots in the study of large number particle systems (e.g. gases), in which microscopic chaos (irregular behavior of individual particles) leads to macroscopic statistical regularity on the entire set of particles. This regularity (characteristic of so-called ergodic systems) is expressed in the stationary density of state distribution reached after a certain time of a system's dynamic evolution. Having considered that, we assume that the initial condition for a dynamic system is a particular density function, and then we study its evolution over time. A book by Lasota and Mackey (1994) is devoted to such approach of dynamic systems studying. Density transformation is determined by a linear operator called the Frobenius–Perron operator. It can be determined for some systems, and for others it is difficult, and then one may attempt to approximate density evolution numerically, for example, by determining a very large set of trajectories. It can be said that we then examine the "average" behavior of the system. This concept was introduced by J.W. Gibbs (see Dorfman 2001, p. 18, 65), who stated that since the initial condition of individual trajectories is never exactly known (with infinite precision), the average behavior

P. J. Mitkowski, *Mathematical Structures of Ergodicity and Chaos in Population Dynamics*, Studies in Systems, Decision and Control 312, https://doi.org/10.1007/978-3-030-57678-3_4

of the entire set of trajectories corresponding to the same equilibrium state of the system can be studied. In physics, large sets of system trajectories are called Gibbs sets (see e.g. Lebowitz and Penrose 1973).

Further in this monograph (Chap. 6), we will numerically approximate the evolution of density for an infinitely dimensional hematological model (6.5). In order to understand well the essence of the density evolution issue, we will first consider it for a simpler system, i.e. for the one-dimensional logistic mapping (4.3). Earlier, however, we will formalize the concept of the Frobenius–Perron operator.

4.1.1   Frobenius–Perron Operator

The following set of concepts was prepared based on the book of Lasota and Mackey (1994, p. 41).

The inverse image of a set $A \subset X$, on which a transformation $S: X \to X$ acts is called a set of all points, which will be in A after one application of S, what can be written

$$S^{-1}(A) = \{x: S(x) \in A\}. \tag{4.1}$$

Definition 4.1 Let $(X, \mathcal{A}, \mu)$ be a measure space. Transformation $S: X \to X$ is measurable if

$$S^{-1}(A) \in \mathcal{A} \quad \text{for all } A \in \mathcal{A}.$$

Definition 4.2 A measurable transformation $S: X \to X$ on a measure space $(X, \mathcal{A}, \mu)$ is nonsingular if $\mu(S^{-1}(A)) = 0$ for all $A \in \mathcal{A}$ such that $\mu(A) = 0$.

Definition 4.3 Let $(X, \mathcal{A}, \mu)$ be a measure space. If $S: X \to X$ is a nonsingular transformation the unique operator $P: L^1 \to L^1$ defined by the following equation:

$$\int_A Pf(x)\mu(dx) = \int_{S^{-1}(A)} f(x)\mu(dx) \quad \text{for } A \in \mathcal{A} \tag{4.2}$$

is called the Frobenius–Perron operator.

Frobenius–Perron operators are special class of Markov operators (see Lasota and Mackey 1994, p. 37 and 41).

4.1.2   Example of Density Evolution Approximation

Before giving an example, let us formalize two terms that will be used further (see Lasota and Mackey 1994, p. 41)

Definition 4.4 Let $(X, \mathcal{A}, \mu)$ be a measure space and let a set $D(X, \mathcal{A}, \mu) = \{f \in L^1(X, \mathcal{A}, \mu): f \geq 0 \text{ and } ||f|| = 1\}$. Every function $f \in D(X, \mathcal{A}, \mu)$ is called density.

Definition 4.5 If $f \in L^1(X, \mathcal{A}, \mu)$ and $f \geq 0$, then the measure

$$\mu_f(A) = \int_A f(x)\mu(dx),$$

is called absolutely continuous with respect to a measure μ and f is called Radon-Nikodym derivative of μ_f with respect to μ. In the particular case when f is a density, we also say, that f is a density of μ_f and that μ_f is normalized measure.

The property stating that a measure ν is absolutely continuous with respect to μ if $\nu(A) = 0$ always, when $\mu(A) = 0$ is frequently given as a definition of absolutely continuous measure. It is also sometimes said that an absolutely continuous measure is one that has a density.

Lat us now consider logistic map

$$S(x) = 4x(1 - x), \quad \text{for } x \in [0, 1] \tag{4.3}$$

transforming closed unit interval $[0, 1]$ onto itself. One can determine (see Lasota and Mackey 1994, p. 7) (see also Ulam 1960, p. 74), that for (4.3) Frobenius–Perron operator has the form

$$Pf(x) = \frac{1}{4\sqrt{1-x}}\left[f\left(\frac{1}{2} - \frac{1}{2}\sqrt{1-x}\right) + f\left(\frac{1}{2} + \frac{1}{2}\sqrt{1-x}\right)\right]. \tag{4.4}$$

Equation (4.4) describes how the map (4.3) transforms given density f to new density Pf. Let us take for example initial density $f(x) \equiv 1$ (see Lasota and Mackey 1994). When we substitute it to the Eq. (4.4) we obtain new density

$$Pf(x) = \frac{1}{2\sqrt{1-x}}, \tag{4.5}$$

after one iteration of the map (4.3). Now substituting it to the Eq. (4.4) we obtain density after second iteration of the map (4.3)

$$P(Pf(x)) = P^2 f(x) = \frac{\sqrt{2}}{8\sqrt{1-x}}\left[\frac{1}{\sqrt{1+\sqrt{1-x}}} + \frac{1}{\sqrt{1-\sqrt{1-x}}}\right]. \tag{4.6}$$

In this way, we can iteratively determine evolution of initial density $f(x) \equiv 1$. In Fig. 4.1a, b, c initial density $f(x) \equiv 1$ and its transformation given by the Eqs. (4.5) and (4.6) are presented. Limit density exists for the map (4.3) and every initial density converge to it in the process of evolution. Ulam and von Neumann (1947) showed

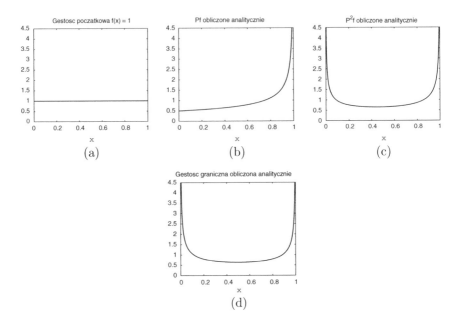

Fig. 4.1 Frobenius–Perron operator corresponding to logistic map (4.3) **a**: initial density $f(x) \equiv 1$, $x \in [0, 1]$ **b**: Pf **c**: $P^2 f$ **d**: limit density (4.7)

(see also Lasota and Mackey 1994, p. 8 and 53) (and Ulam 1960, p. 74), that this limit density has a form

$$f_{gr}(x) = \frac{1}{\pi \sqrt{x(1-x)}}. \tag{4.7}$$

It is shown in Fig. 4.1d.

In the case of the map (4.3), the density evolution can be exactly determined. In the further part of this book (Chap. 6), we will consider a system for which it is not known whether it can be done, therefore, we will be willing to approximate the evolution of density numerically. At the moment, we are approximating the evolution of density for the mapping (4.3) numerically to verify what results will be given by the numerical method, which we will use in comparison with the exact solutions. For approximation of density evolution, we will use the simplest method, i.e. we will calculate a large set of map (4.3) trajectories and determine histograms counting the number of trajectory points in subsets of the entire space [0, 1] for subsequent map iterations. Histograms will then be normalized to histograms with a unit surface area, thus obtaining an approximation of the density function. Figures 4.2 and 4.3 show the result of such an operation. In the left column, we approximate the evolution of the homogeneous distribution, and in the right, the normal distribution. Distributions were given for 5000 starting points. It can, therefore, be seen that for a homogeneous distribution, the numerical results obtained are very similar to the exact results. Indeed, histograms for the first three iterations (Fig. 4.2a, c, d) very closely approximate the first three

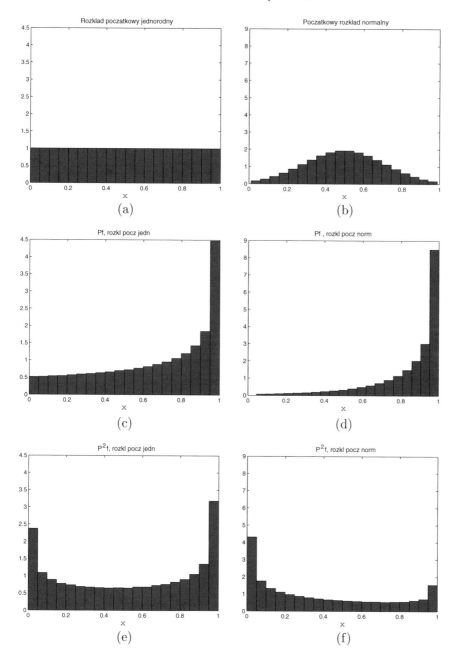

Fig. 4.2 The left column downward contains the evolution of a homogeneous distribution of 5000 points under the action of the logistic map (4.3). The right column downward contains the evolution of the normal distribution under the action of the logistic map (4.3). Herein, (**a, b**): initial distributions, (**c, d**): distributions after the first map iteration, (**e, f**): after the second iteration

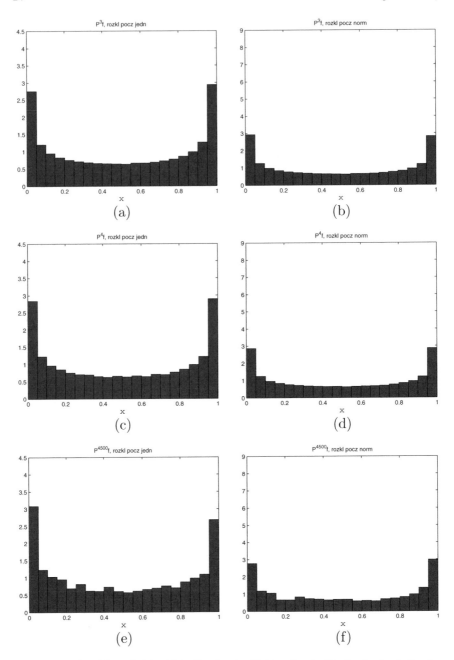

Fig. 4.3 Continued from the Fig. 4.2 (**a, b**): after the third iteration (**c, d**): after the fourth iteration (**e, f**): after 4500 iterations

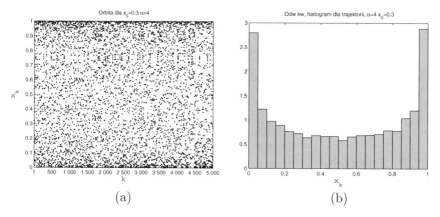

Fig. 4.4 a: The logistic map (4.3) trajectory for the initial point $x_0 = 0.3$ after 5000 map iterations
b: histogram for points along this trajectory

iterations of the Frobenius–Perron operator (4.4) for constant initial density $f(x) \equiv 1$
(Fig. 4.1a, b, c). For both distributions (homogeneous and normal), the histograms
tend to be shaped very similarly to the limit density function (4.7).

When drawing the histogram according to the same principle as before, but along
a single long trajectory of the map (4.3) (see Fig. 4.4), the result will be very similar
to the limit density function (4.7). This is a typical property of ergodic systems. We
are going to discuss this in detail in the next chapter.

4.2   Chaos for Measure-Preserving Systems

One of the most fundamental concepts in ergodic theory is the concept of invariant
measure (see Lasota and Mackey 1994; Fomin et al. 1987; Bronsztejn et al. 2004;
Rudnicki 2004; Dawidowicz 2007; Collet and Eckamnn 1980; Parry 1981; Collet
2005), which is a consequence of Liouville's theorem (see e.g. Szlenk 1982; Landau
and Lifszyc 2007; Arnold 1989; Nadzieja 1996; Dorfman 2001). Maps (or flows)
maintaining measure show three main levels of behavior irregularities (from lowest
to highest level): ergodicity, mixing and exactness. Light mixing, mild mixing and
weak mixing can be still distinguished between the ergodicity and mixing (Lasota and
Mackey 1994; Silva 2010). The so-called K-flows (or K-property, K-automorphism)
are determined at a level similar to exactness (see Rudnicki 1985a, b, 2004) and
(Lasota and Mackey 1994). In this book, we will only consider ergodicity and mixing.
We will now formalize these concepts and provide examples of low-dimensional
ergodic and mixing maps to understand the characteristic behavior associated with
these properties.

By $\{S_t\}_{t \geq 0}$ let us denote semidynamic system or semiflow on the metric space X,
i.e.

1. $S_0(x) = x$ for all $x \in X$,
2. $S_t(S_{t'}(x)) = S_{t+t'}(x)$ for all $x \in X$, and $t, t' \in R^+$,
3. $S: X \times R^+ \to X$ is a continuous function (t, x).

By a measure on X we mean any probabilistic measure defined on a σ-algebra of Borel subsets of set X. A measure μ is called invariant under a semiflow $\{S_t\}_{t \geq 0}$ if $\mu(A) = \mu(S^{-t}(A))$ for each $t \geq 0$ and all $A \in \mathcal{B}$. $S^{-t}(A) := (S^t)^{-1}(A)$.

4.2.1 Ergodicity

By (S, μ) let us denote a semiflow $\{S_t\}_{t \geq 0}$ with an invariant measure μ. Semiflow (S, μ) is ergodic (we also say, that a measure is ergodic) if a measure $\mu(A)$ of any invariant set A equals 0 or 1. Let us now consider two simple examples.

Example 4.1 Let $S: [0, 2\pi) \to [0, 2\pi)$ be a transformation of rotation on a circle of radius 1 by an angle ϕ (see Lasota and Mackey 1994; Bronsztejn et al. 2004; Devaney 1987; Dorfman 2001)

$$S(x) = x + \phi \qquad (\text{mod } 2\pi). \tag{4.8}$$

When the expression $\phi/2\pi$ is rational, we can find invariant sets, which measure differs from 0 or 1, therefore, S is not ergodic (see Fig. 4.5a). However when this expression is irrational, then S is ergodic (proof can be found in (Lasota and Mackey 1994, p. 75) or in a different perspective in (Devaney 1987, p. 21)). If we take, e.g., $\phi = \sqrt{2}$ and pick an arbitrary point on a circle, we can observe, that a subsequent iterations of this point under the action of S densely fill available space i.e. circle (see Fig. 4.5b, c).

Example 4.2 To understand better typical features of ergodic behavior, let us consider the following transformation (see Lasota and Mackey 1994, p. 68)

$$S(x, y) = (\sqrt{2} + x, \sqrt{3} + y) \qquad (\text{mod } 1). \tag{4.9}$$

It is an extension of the rotation transformation (4.8) from the previous example on a space $[0, 1] \times [0, 1] \to [0, 1] \times [0, 1]$. In Fig. 4.6 the result of the map (4.9) action on a set of 10^3 points distributed randomly in the area $[0, 0.1] \times [0, 0.1]$ can be seen. The transformation moves the initial area and does not spread points all over the available space. When we measure the Euclidean distance between two closely placed points from the initial set, we notice that it is constant in each iteration (see Fig. 4.7). Indeed the "popular" criterion of chaos, i.e. sensitivity to a small change in the initial condition is not the property of ergodic systems. Their property however is that they have dense trajectories. We will formulate this fact precisely a little further.

The transformation (4.9) can also be extended to three dimensions, e.g. as follows

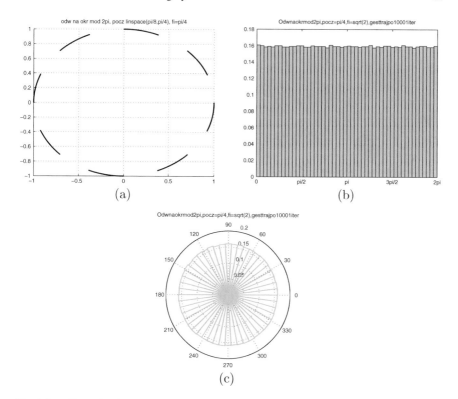

Fig. 4.5 a: Exemplary invariant set for $\phi = \pi/4$ **b**: Normalized (to probability density) histogram showing, that a point under the action of the transformation (4.8) at $\phi = \sqrt{2}$ densely fills a circle **c**: the same normalized histogram, but in the form of a circle (bars inside a circle)

$$S(x, y, z) = (\sqrt{2} + x, \sqrt{3} + y, \sqrt{5} + z) \quad (\text{mod } 1). \qquad (4.10)$$

In Fig. 4.8 we can see how the ergodic system behaves in three-dimensional space.

One of the most important theorems in the ergodic theory is the Birkhoff individual ergodic theorem (Birkhoff 1931a,b; Birkhoff and Koopman 1932; Lasota and Mackey 1994; Fomin et al. 1987; Szlenk 1982; Dawidowicz 2007; Nadzieja 1996; Górnicki 2001; Dorfman 2001). Here we will give a popular extension of this theorem (see Lasota and Mackey 1994, p. 64), (and also Fomin et al. 1987, p. 46). Let us be reminded that by (S, μ) we denote semiflow $\{S_t\}_{t \geq 0}$ with the invariant measure μ.

Theorem 4.1 (Extension of the Birkhoff theorem) *Let (S, μ) be ergodic. Then for each μ-integrable function $f : X \to R$, the average of f along the trajectory of S is equal almost everywhere to the average of f over the space X; that is*

$$\lim_{T \to \infty} \frac{1}{T} \int_0^T f(S_t(x)) \mathrm{d}t = \frac{1}{\mu(X)} \int_X f(x) \mu(\mathrm{d}x) \quad \text{a.e.} \qquad (4.11)$$

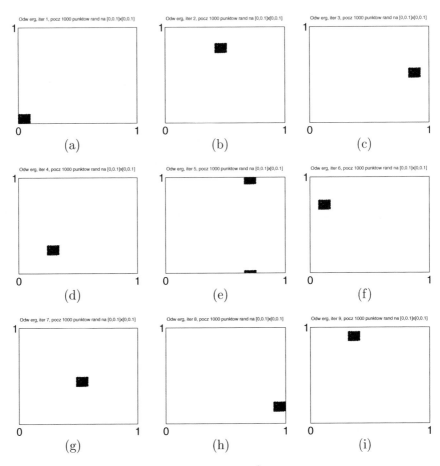

Fig. 4.6 Ergodic transformation (4.9) acting on a set of 10^3 points distributed randomly in $[0, 0.1] \times [0, 0.1]$

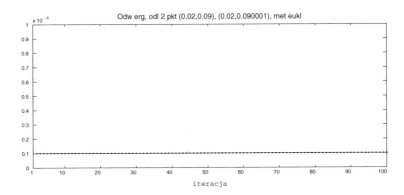

Fig. 4.7 Euclidean metric between two arbitrarily selected close points on which the transformation (4.9) acts

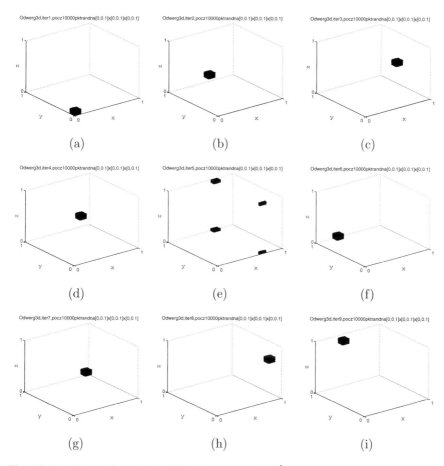

Fig. 4.8 Ergodic transformation (4.10) acting on a set of 10^4 distributed randomly in $[0, 0.1] \times [0, 0.1] \times [0, 0.1]$

If we substitute $f = \mathbf{1}_A$ in Eq. (4.11) (see Lasota and Mackey 1994; Rudnicki 2004; Dawidowicz 2007) then the left-hand side of the Eq. (4.11) is the average time of visiting the set A, and the right-hand side is $\mu(A)$. This corresponds to ergodicity in the sense of Boltzmann, what can be briefly expressed as follows: the average time of the physical system particle staying in a given area is proportional to its natural probabilistic measure (Dawidowicz 2007; Dorfman 2001; Nadzieja 1996; Górnicki 2001; Birkhoff and Koopman 1932; Lebowitz and Penrose 1973).

We could see that ergodicity in its "pure" form does not have to be associated with unpredictable behavior. The invariant and ergodic measure must, thus, have additional properties to be interesting from a dynamic point of view. Very generally speaking, it should be non-trivial, i.e. e.g. it should not be focused on a single point.

While the conditions of Birkhoff theorem are met in a single point (the average over a set set is equal to the average along single trajectories), still this is not a very interesting case. That is because one assume that, to make the conclusions from this theorem interesting, the support of the measure should be fairly large set (see e.g. Bronsztejn et al. 2004, p. 891). According to our knowledge, there are two approaches to this issue. Both seem to be similar in terms of their main ideas, but in the literature they appear separately. One is the theory formulated by Prodi (1960) (as well as by Foias 1973), which states that stationary turbulences occur when the flow has a non-trivial invariant and ergodic measure. This theory was strongly developed by Lasota (1979, 1981) (see also Lasota and Yorke 1977; Lasota and Myjak 2002; Lasota and Szarek 2004) and further by Rudnicki (1985a, 1988, 2009), (see also Myjak and Rudnicki 2002) and Dawidowicz (1992a, b) (see also Dawidowicz et al. 2007). Another approach uses the concept of SRB measures (Sinai, Ruelle and Bowen) (see e.g. Bronsztejn et al. 2004; Dorfman 2001; Taylor 2004; Tucker 1999).

To summarize the considerations about ergodic systems, let us formalize an important conclusion. Let us consider positive μ measures on all non-empty open sub-sets X (see Rudnicki 2004, p. 727, Proposition 1). If semiflow (S, μ) is ergodic, then for μ-a.a. x, trajectory $\vartheta(x) = \{S^t(x): \ t \geq 0\}$ is dense. Thus, having considered the foregoing, dense trajectory is a property of ergodic systems.

4.2.2 Mixing

Let us now consider the concept of **mixing**, which represents a higher level of behavior irregularity than ergodicity. Physical literature states that the concept of a **mixing system** was introduced by J.W. Gibbs (see Dorfman 2001, p. 18, 65). Semiflow (S, μ) is mixing (see e.g. Lasota and Mackey 1994; Rudnicki 2004; Bronsztejn et al. 2004) if

$$\lim_{t \to \infty} \mu(A \cap S_t^{-1}(B)) = \mu(A)\mu(B)) \quad \text{for all } A, B \in \mathcal{B}. \tag{4.12}$$

This means that the fraction of points that for $t = 0$ are in A and for large t are in B is given as a product of any A and B sets measures in X. Mixing systems are also ergodic.

Example 4.3 Let us consider mixing mapping (see Lasota and Mackey 1994, p. 57, 65, 68)

$$S(x, y) = (x + y, x + 2y) \quad (\text{mod } 1). \tag{4.13}$$

It is an example of Anosov diffeormorphism (Anosov 1963) (see also Bronsztejn et al. 2004, p. 903). In Fig. 4.9 the first few iterations of a mapping (4.13) acting on a set of 10^3 points distributed randomly over the area $[0, 0.1] \times [0, 0.1]$ are presented. The points are scattered all over the available space and then the mapping literally 'mixes' them throughout the entire space. The Euclidean distance between any two points starting very close to each other first increases rapidly and then oscillates irregularly (see Fig. 4.10). Thus, there is a clear difference in the behavior of the mixing system and the ergodic system discussed earlier. A typical property of mixing systems is sensitivity to a small change in initial conditions.

Similarly, to the ergodic system (4.9), we can extend the mixing mapping (4.13) to three dimensions, e.g. to the following mapping:

$$S(x, y, z) = (x + y + z, x + 2y + z, x + y + 3z) \qquad (\text{mod } 1). \qquad (4.14)$$

Figure 4.11 shows first few iterations of the mapping (4.14).

We can say more about the chaotic nature of mixing systems. First, however, let us recall the following definition (Auslander and Yorke 1980), (see also Rudnicki 2004).

Definition 4.6 A flow is chaotic in the sense of Auslander and Yorke if

1. there exists a dense trajectory,
2. each trajectory is unstable.

Instability we understand in the following sense. There exists a constant $\eta > 0$, such that for each point $x \in X$ and for each $\epsilon > 0$ there exists a point $y \in B(x, \epsilon)$ and $t > 0$, such that $\rho(S_t(x), S_t(y)) > \eta$ where ρ is a metric in X and $B(x, \epsilon)$ is a sphere in X with center x and radius $r > 0$. Instability can be defined as sensitivity to small changes in initial conditions, what is a "popular" criterion of chaos. When we consider again positive μ measures on all non-empty open subsets of X (see Rudnicki 2004, p. 727, Proposition 1), we can state that if semiflow (S, μ) is mixing, then semiflow $\{S_t\}_{t \geq 0}$ is chaotic in the sense of Auslander and Yorke.

Example 4.4 An important property of mixing systems is that in a result of dynamic evolution they reach a state of statistical equilibrium in the entire space (in which they operate) and in its subspaces. The convergence to stationary distribution in subspaces is faster than in the entire space. We can observe this on the example of the mapping (4.13) considered above. In Figs. 4.12 and 4.13, we can observe how the initial normal distribution of 10^5 points, on which mapping (4.13) acts, converge in subsequent iterations on the intervals [0, 1] of the x and y axes (on subspaces of the entire space [0, 1] × [0, 1]) to a stationary homogeneous distribution that is already achieved in the fourth iteration. Normal distribution in the entire space [0, 1] × [0, 1] (see Fig. 4.14) also converges to stationary homogeneous distribution but is achieved in the seventh iteration.

Let us note one more fact. The presented histograms of evolution of distributions of a large set of points were built by counting the number of these points in given subintervals of the section [0, 1] and the surface [0, 1] × [0, 1]. When defining each point in the Euclidean norm now, and drawing up a histogram counting the numbers of the values for these norms, we will notice that it also converges to stationary distribution (see Fig. 4.15), but with a different shape than in previous cases. When checking the typical ergodicity property (the mixing system is also ergodic, as we mentioned

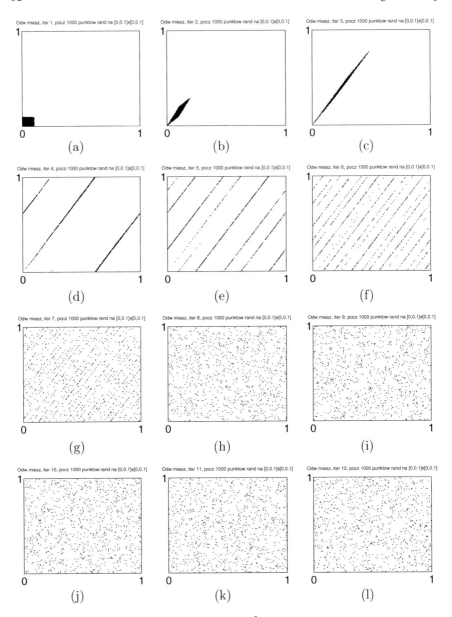

Fig. 4.9 Mixing mapping (4.13) acting on a set of 10^3 points distributed randomly in $[0, 0.1] \times [0, 0.1]$

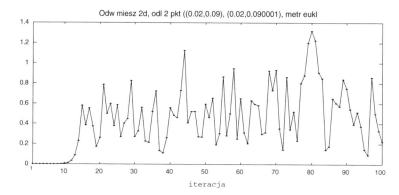

Fig. 4.10 Euclidean metric between two arbitrarily selected close points, on which the mapping (4.13) acts

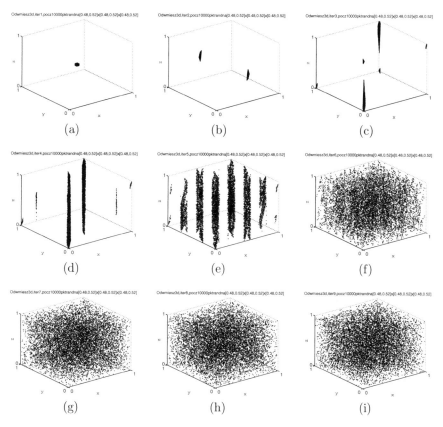

Fig. 4.11 Mixing mapping (4.14) acting on a set of 10^4 points distributed randomly in $[0.48, 0.52] \times [0.48, 0.52] \times [0.48, 0.52]$

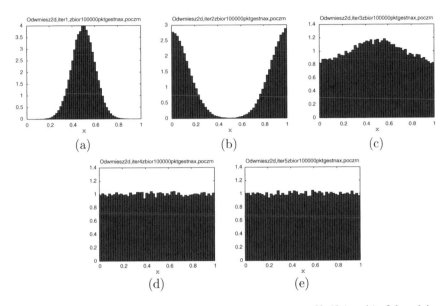

Fig. 4.12 Convergence to a stationary distribution on the subspace $[0, 1]$ (x-axis) of the mixing mapping (4.13) acting on 10^5 points with the initial normal distribution

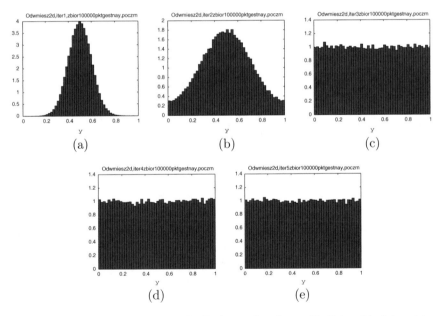

Fig. 4.13 Convergence to a stationary distribution on the subspace $[0, 1]$ (y-axis) of the mixing mapping (4.13) acting on 10^5 points with the initial normal distribution

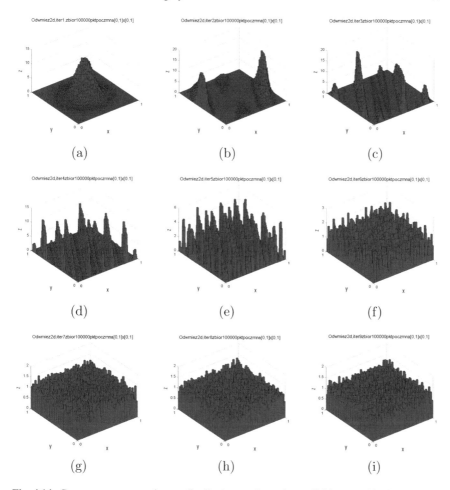

Fig. 4.14 Convergence to a stationary distribution on the entire available space $[0, 1] \times [0, 1]$ of the mixing mapping (4.13) acting on 10^5 points with the initial normal distribution

earlier), the distributions along individual trajectories in the respective spaces will be the same as the distributions for large sets of points (compare Fig. 4.16a, b, c, d). This observation may have important practical consequences in the experimental, numerical study of more complex systems operating in abstract spaces. One should be aware that equilibrium distribution functions may have various shapes depending on the assumed space for study.

Example 4.5 Let us analyze another characteristic property of mixing systems. In order to do so, let us consider the correlation coefficient in the following form (see de Larminat and Thomas 1983):

$$\gamma_{xy}(\tau) = \frac{c_{xy}(\tau)}{\sigma_x \sigma y}, \tau = 0, 1, 2, \ldots \tag{4.15}$$

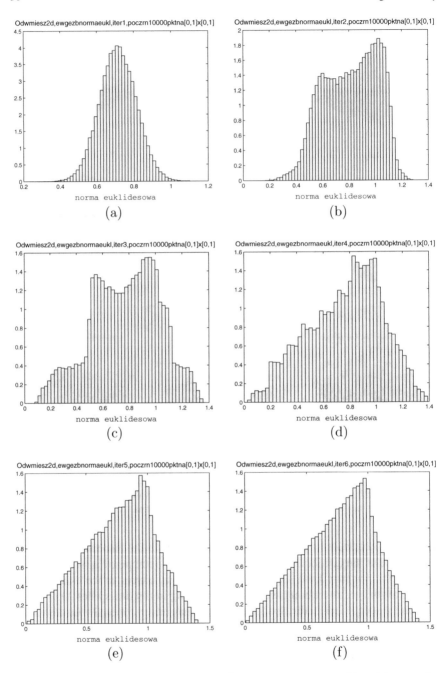

Fig. 4.15 Evolution of normal distribution of 10^5 points under the action of the mixing mapping (4.13) in the space $[0, 1] \times [0, 1]$ with Euclidean norm

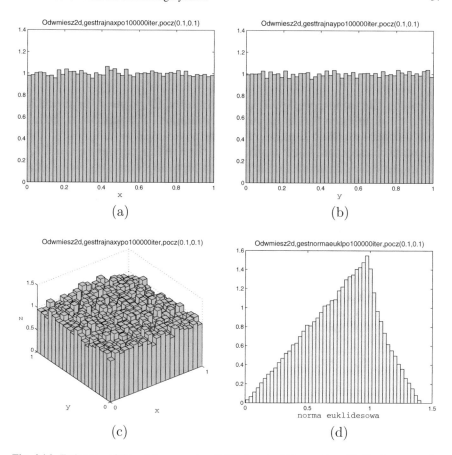

Fig. 4.16 Trajectory of the mixing mapping (4.13) densely filling sections [0, 1] of axes **a**: x **b**: y and **c**: entire space $[0, 1] \times [0, 1]$ **d**: distribution along this trajectory in the space $[0, 1] \times [0, 1]$ with Euclidean norm. All histograms are calculated for 10^5 of trajectory points

where

$$c_{xy}(\tau) = \lim_{N \to \infty} \frac{1}{N} \sum_{i=1}^{N} (x_i - x_0)(y_{i+\tau} - y_0(\tau)), \tag{4.16}$$

$$x_0 = \lim_{N \to \infty} \frac{1}{N} \sum_{i=1}^{N} x_i, \quad y_0(\tau) = \lim_{N \to \infty} \frac{1}{N} \sum_{i=1}^{N} y_{i+\tau} \tag{4.17}$$

and

$$\sigma_x = \sqrt{\lim_{N \to \infty} \frac{1}{N} \sum_{i=1}^{N} (x_i - x_0)^2}, \tag{4.18}$$

$$\sigma_y = \sqrt{\lim_{N\to\infty} \frac{1}{N} \sum_{i=1}^{N} (y_{i+\tau} - y_0(\tau))^2}. \qquad (4.19)$$

Let us consider the mapping (4.13) acting on a set of 10^4 points. After several iterations the distribution for the set reaches equilibrium. For such set in statistical stationary state we take a sequence of Euclidean norms (i.e. x_i in Eq. (4.15)), therefore we have a sequence of 10^4 numbers $y_{i+\tau}$ for $\tau = 0$ i.e. x_i, and for $\tau = 1, 2, \ldots$ is a sequence for next iterations. Therefore the Eq. (4.15) is the correlation function between the selected initial sequence for $\tau = 0$, and a sequence for following iterations. The result of calculating this function is shown in Fig. 4.17a. It can be seen that the correlation for a set of points decays to a value close to 0 already in the second iteration (the first iteration is the correlation of the initial sequence with itself, so it is 1). When drawing the spread of points after several iterations (see Fig. 4.17d), we notice that the points are not correlated either linearly or in any other way. When we calculate correlations in the same way but for a sequence of values of single long trajectory and a sequence of its time shifts, we obtain (see Fig. 4.9b) a correlation function of the same character as for a large set of points. This rapid decay of correlation is typical for mixing systems (see Bronsztejn et al. 2004; Rudnicki 2004, 1988).

4.2.3 Turbulence in Mixing Systems

Decay of correlation for single trajectories and their time shifts defines turbulent trajectories, which are a property of mixing systems. Formalizing, let us consider the following definition (see Bass 1974; Rudnicki 2004)

Definition 4.7 A trajectory $\vartheta(x) = \{S^t(x) : t \geq 0\}$ of a point $x \in X$ is called turbulent in the sense of Bass if there exists a x_0 and a function γ such that:

1. $\lim_{T\to\infty} \frac{1}{T} \int_0^T S^t(x)dt = x_0$,
2. $\lim_{T\to\infty} \frac{1}{T} \int_0^T (S^t(x) - x_0)(S^{t+\tau}(x) - x_0)dt = \gamma(\tau)$,
3. $\gamma(0) \neq 0$ i $\lim_{\tau\to\infty} \gamma(\tau) = 0$.

One can also formulate the following statement (see Rudnicki 2004, p. 727, Proposition 2), (Rudnicki 1988, p. 17, Corollary 2), (Myjak and Rudnicki 2002). Let us assume, that measure μ is not supported on a single point. If a semiflow (S, μ) is mixing and

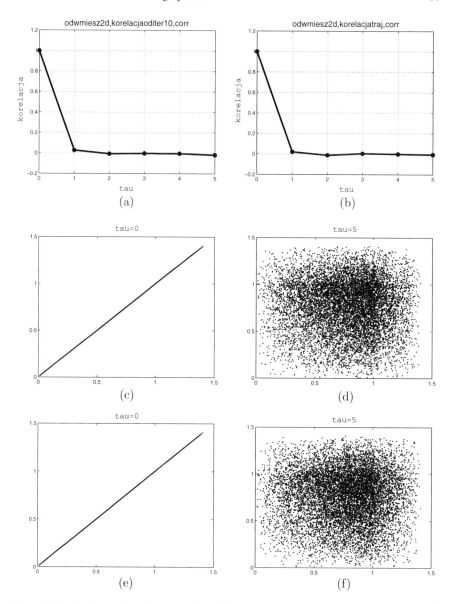

Fig. 4.17 Rapid decay of correlation for the mixing mapping (4.13) **a**: correlation for a set of 10^4 points **b**: correlation for single trajectory and its time shift (**c** and **d**): spread for set of trajectories and time shifts, respectively, $\tau = 0$ and $\tau = 5$ (**e** and **f**): spread for single trajectory and its time shifts, respectively, $\tau = 0$ i $\tau = 5$

Fig. 4.18 The map (4.21), for exemplary values $a = \frac{1}{2}$ and $a = \frac{3}{4}$ satisfies the condition of the Ulam hypothesis, however at $a = \frac{1}{4}$ it does not satisfy it

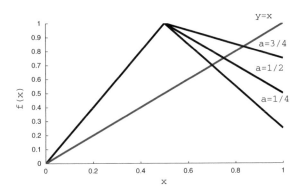

$$\int_X ||x||^2 \mu(dx) < \infty, \tag{4.20}$$

then for almost all x trajectory $\vartheta(x)$ is turbulent in the sense of Bass.

Thus, in mixing systems, microscopic chaos (for single trajectories) can be expressed in the language of turbulence theory. This microscopic chaos leads to statistical regularity of mixing systems, which is their interesting property observed in many real processes.

4.3 Ulam Hypothesis

Ulam (1960, p. 74) formulated the problem of existence of non-trivial invariant measure for the map $f(x)$ of section $[0, 1]$ into itself defined by (sufficiently) simple function (e.g. broken line or polynomial), the graph of which does not intersect the line $y = x$ with a angle tangent of an absolute value less than 1. E.g. the map

$$f(x) = \begin{cases} 2x, & 0 \le x \le \frac{1}{2} \\ (2 - a) + 2(a - 1)x, & \frac{1}{2} \le x \le 1, \end{cases} \tag{4.21}$$

at $a = \frac{1}{2}$ or $a = \frac{3}{4}$ (see Fig. 4.18) fulfils this assumption, while at $a = \frac{1}{4}$ it does not. Lasota and Yorke (1973) gave the solution to this problem. Ulam hypothesis is also mentioned by (Myjak 2008, p. 376).

We refer to this issue because it seems to have an interesting association with the problems brought up in this monograph. We will discuss this in detail in Chap. 6.

Chapter 5
The Lasota–Ważewska Equation

Polish scientists, the hematologist Maria Ważewska-Czyżewska and the mathematician Andrzej Lasota, formulated a mathematical model of the red blood cell system dynamics, known today as the Lasota–Ważewska equation, in an article published in *Applied Mathematics* (annals of the Polish Mathematical Society) in the year 1976. Their main goal was to provide a model that, with the possible low number of constants having a specified biological meaning, would allow to predict changes in the number of blood cells in the bloodstream. The basic idea was to associate some known equation of population dynamics taking into account the age structure of the population, with a properly constructed feedback in the form of an integral equation with a delayed argument. By doing so, they obtained a model that in many cases provides solutions corresponding to biological and medical experimental data, and which became the inspiration for studies on many issues in the field of pure mathematics and biomathematics (Rudnicki 2007).

5.1 McKendrick–Von Foerster Equation

The above-mentioned equation used by M. Ważewska-Czyżewska and A. Lasota in their model is the equation that was first proposed by Sharpe and Lotka (1911) (see Rudnicki 2014, p. 216), then by McKendrick (1926) and Von Foerster (1959). This is an equation describing the age distribution of the population over time. A. G. McKendrick used it to model the human population attacked by epidemics, while H. Von Foerster considered it in the context of hematological experimental data. In literature, it is called the "McKendrick" equation, or often the "Von Foerster" equation. We are going to use the name "McKendrick–Von Foerster" equation. It is one of the oldest of so-called structural models (Rudnicki 2014, p. 216), (Murray 2006, p. 39), i.e. taking into account the natural structure for the studied population, e.g. age, spatial, maturity, size, etc.

© The Editor(s) (if applicable) and The Author(s), under exclusive license
to Springer Nature Switzerland AG 2021
P. J. Mitkowski, *Mathematical Structures of Ergodicity and Chaos
in Population Dynamics*, Studies in Systems, Decision and Control 312,
https://doi.org/10.1007/978-3-030-57678-3_5

We have mentioned that A. G. McKendrick used the equation to model dynamics of human population attacked by epidemics, in particular he was studied the variability of the population age distribution. Humankind has been affected significantly by infectious diseases throughout its known history (Thieme 2003, p. 283), (Rudnicki 2014, p. 84). Around 430 B.C. unknown epidemic contribute to the fall of Athens. The Plague of Justynian attacked the Byzantine and Sasanian empires, as well as the port cities around the whole Mediterranean See in 541-542 A.D. and returning until 750, killing around 25–100 millions people. The Black Death caused decease of up to 75–200 millions people in Eurasia and North Africa. In 1347–1351 around 1/4 of Europe population has fallen because of this pandemic. The history of mankind know more such examples. Infectious diseases (HIV, Ebola, Malaria, lassa, hanta, dengue viruses and yellow fever) causes still a continuing threat. Currently we are witnessing coronavirus SARS-CoV-2 pandemic. We can observe that population dynamics models are now intensively applied in order to predict this pandemic behaviour (see e.g. Peng et al. 2020; Luo et al. 2020; Khan and Atangana 2020; Kucharski et al. 2020; Prem et al. 2020; Cherniha and Davydovych 2020).

Knowledge of the structure and principles of the McKendrick–Von Foerster equation is important for understanding the models analyzed later in this book, therefore, we are going to derive this equation now (see Ważewska-Czyżewska and Lasota 1976; Rudnicki 2014). It is worth referring to the source papers of Sharpe and Lotka (1911), McKendirck (1926), and Von Foerster (1959).

Let us denote the density function of the age distribution for red blood cells population by $n(t, a)$, satisfying the condition

$$\int_0^\infty n(t, a)da = N(t),$$ (5.1)

where $N(t)$ is a total number of blood cells in the bloodstream at time t. In other words, the function $n(t, a)$ determines the number of blood cells in age a for a given time t. Blood cells that were at age a at time t, are at age $a + \Delta t$ at time $t + \Delta t$, therefore, the difference

$$n(t, a) - n(t + \Delta t, a + \Delta t),$$ (5.2)

gives the number of blood cells at age a that died within the time interval $(t, t + \Delta t)$. The limit

$$i(t, a) = \lim_{\Delta t \to \infty} \frac{n(t, a) - n(t + \Delta t, a + \Delta t)}{\Delta t}$$ (5.3)

denotes the **destruction intensity** of blood cells at age a and at time t. The quotient

$$\lambda(t, a) = \frac{i(t, a)}{n(t, a)}$$ (5.4)

is called **coefficient of destruction** and it determines the probability that a blood cell which in time t is at age a will die within a period of time $(t, t + \Delta t)$. From (5.3) and (5.4) we obtain

$$\lim_{\Delta t \to \infty} \frac{n(t, a) - n(t + \Delta t, a + \Delta t)}{\Delta t} = \lambda(t, a)n(t, a). \tag{5.5}$$

Assuming now that there are partial derivatives of the function $n(t, a)$, the formula (5.5) can be written in a form of partial equation

$$\frac{\partial n(t, a)}{\partial t} + \frac{\partial n(t, a)}{\partial a} = -\lambda(t, a)n(t, a). \tag{5.6}$$

The Eq. (5.6) is the McKendrick–Von Foerster equation we are looking for. As can be seen from the presented formulation, it is a consequence of only the destruction coefficient λ. Von Foerster (1959, p. 393) characterizes this coefficient as the sum of "intrinsic loss" caused by aging of population elements and "environmental loss" resulting from their interaction with the external environment.

To solve the Eq. (5.6), one must know the initial condition

$$n(0, a) = v(a), \tag{5.7}$$

i.e. the age distribution $v(a)$ at time $t = 0$ and the boundary condition

$$n(t, 0) = p(t), \tag{5.8}$$

interpreted as the blood cell production $p(t)$ at time t (see Ważewska-Czyżewska and Lasota 1976, p. 25). The Eq. (5.6) with conditions (5.7) and (5.8) is an example of "self-regulating" mechanism, as McKendrick emphasized in his work (1926, p. 124).

5.1.1 Method of Characteristics

Equation (5.6) can be solved using method of characteristics (Murray 2006, p. 41), (Rudnicki 2014, p. 209), (Pelczar and Szarski 1987, p. 253). Family of characteristics are solutions of equation

$$\dot{a} = 1, \tag{5.9}$$

so they are straight lines (see Fig. 5.1)

$$a = \begin{cases} t + a_0 \text{ dla } a > t, \\ t - t_0 \text{ dla } a < t, \end{cases} \tag{5.10}$$

Fig. 5.1 Characteristics for
the McKendrick–Von
Foerster equation (5.6)

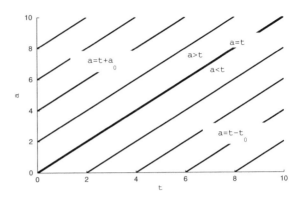

where a_0 is the individual's initial age at time $t = 0$, and t_0 is the time (moment) at which the individual was born. Solution of the partial equation (5.6) is "strew" on characteristics by solutions of the equation

$$\dot{n} = -\lambda \cdot n. \tag{5.11}$$

By making an appropriate calculations, one can write this in the following form (see e.g. Murray 2006, p. 41)

$$n(t, a) = v(a - t) \exp\{-\int_{a-t}^{a} \lambda(r) dr\}, \qquad a \geq t \tag{5.12}$$

$$n(t, a) = f(t - a) \exp\{-\int_{0}^{a} \lambda(r) dr\}, \qquad a \leq t \tag{5.13}$$

5.1.2   Method of Straight Lines

For Eq. (5.6), which is a linear partial differential equation, it is possible to determine the exact solution. This is not always easy in the case of more complex nonlinear partial equations. In addition, when the characteristics are not straight lines, but take the form of some non-linear functions, the use of formulas obtained by the characteristics method for numerical calculations can be troublesome. We will now present a method that provides approximate solutions, sometimes called the "method of straight lines". It is based on converting a partial differential equation into a system of ordinary differential equations. The number of these equations is equal to the number of discretization points of the spatial variable (i.e. the variable a in the case of Eq. (5.6)). An important advantage of the method of straight lines is its easiness to use in numerical calculations. When we replace Eq. (5.6) with a system of a finite number of ordinary differential equations, then from the infinite-dimensional system we obtain its finite-dimensional approximation.

We will now use the method of straight lines to solve the Eq. (5.6) numerically.

5.1.2.1 Preparation of Numerical Calculations

We discretize a spatial variable (see Fig. 5.2)

$$a_i = i \cdot s, \qquad i = 0, 1, 2, 3, \ldots, M \to +\infty$$

and denote

$$n(t, a_i) = n_i(t), \qquad n(t, 0) = n_0(t) = p(t).$$

We approximate the derivative of the function $n(t, a_i)$ over the variable a

$$\frac{\partial n(t, a_i)}{\partial a} \simeq \frac{n_i(t) - n_{i-1}(t)}{s}$$

and from the Eq. (5.6) we obtain

$$\dot{n}_i(t) + \frac{n_i(t) - n_{i-1}(t)}{s} = -\lambda_i(t) \cdot n_i(t), \qquad i = 1, 2, \ldots, M \to +\infty$$

and further

$$\dot{n}_i(t) = -(\lambda_i(t) + \frac{1}{s})n_i(t) + \frac{1}{s}n_{i-1}(t), \qquad i = 1, 2, \ldots, M \to +\infty. \qquad (5.14)$$

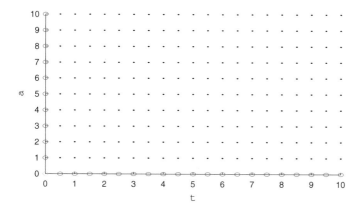

Fig. 5.2 Scheme of the method of straight lines. For given, in points of discretization, values of initial condition (red circles on a axis) and for given boundary condition (blue circles on t axis) we calculate approximate values of $n(t, a)$ in points marked with black dots

We write the Eq. (5.14) in the following form

$$
\begin{cases}
\dot{n}_1(t) = -(\lambda_1(t) + \frac{1}{s})n_1(t) + \frac{1}{s}n_0(t), \qquad n_0(t) = p(t) \\
\dot{n}_2(t) = -(\lambda_2(t) + \frac{1}{s})n_2(t) + \frac{1}{s}n_1(t) \\
\dot{n}_3(t) = -(\lambda_3(t) + \frac{1}{s})n_3(t) + \frac{1}{s}n_2(t) \\
\vdots \\
\dot{n}_M(t) = -(\lambda_M(t) + \frac{1}{s})n_M(t) + \frac{1}{s}n_{M-1}(t)
\end{cases}
$$

and go to the matrix notation

$$
\begin{bmatrix} \dot{n}_1(t) \\ \dot{n}_2(t) \\ \dot{n}_3(t) \\ \vdots \\ \dot{n}_M(t) \end{bmatrix}
=
\begin{bmatrix}
-\lambda_1 - \frac{1}{s} & 0 & 0 & 0 & 0 \\
\frac{1}{s} & -\lambda_2 - \frac{1}{s} & 0 & 0 & 0 \\
0 & \frac{1}{s} & -\lambda_3 - \frac{1}{s} & 0 & 0 \\
0 & 0 & \ddots & \ddots & 0 \\
0 & 0 & 0 & \frac{1}{s} & -\lambda_M - \frac{1}{s}
\end{bmatrix}
\begin{bmatrix} n_1(t) \\ n_2(t) \\ n_3(t) \\ \vdots \\ n_M(t) \end{bmatrix}
+ \frac{1}{s}
\begin{bmatrix} n_0(t) \\ 0 \\ 0 \\ \vdots \\ 0 \end{bmatrix},
$$

co zapisujemy

$$
\dot{n}(t) = A\,n(t) + B\,p(t), \qquad n(t) = \begin{bmatrix} n_1(t) \\ n_2(t) \\ n_3(t) \\ \vdots \\ n_M(t) \end{bmatrix} \in R^M, \tag{5.15}
$$

gdzie

$$
A =
\begin{bmatrix}
-\lambda_1 - \frac{1}{s} & 0 & 0 & 0 & 0 \\
\frac{1}{s} & -\lambda_2 - \frac{1}{s} & 0 & 0 & 0 \\
0 & \frac{1}{s} & -\lambda_3 - \frac{1}{s} & 0 & 0 \\
0 & 0 & \ddots & \ddots & 0 \\
0 & 0 & 0 & \frac{1}{s} & -\lambda_M - \frac{1}{s}
\end{bmatrix}_{M \times M},
$$

$$
\lambda_i(t) = \lambda(t, a_i), \qquad B = \begin{bmatrix} \frac{1}{s} \\ 0 \\ 0 \\ \vdots \\ 0 \end{bmatrix} \in R^M.
$$

Now we discretize the variable t (time)

$$
t = t_k = k \cdot \tau, \qquad k = 0, 1, 2, 3, \ldots, N \to +\infty
$$

and denote

$$n(t_k) = n[k] = \begin{bmatrix} n_1[k] \\ n_2[k] \\ \vdots \\ n_M[k] \end{bmatrix} \in R^M, \qquad \lambda_i(t_k) = \lambda_i(k \cdot \tau).$$

We approximate the derivative of n over the variable t

$$\dot{n}(t) \cong \frac{n(t) - n(t - \tau)}{\tau} = \frac{n[k] - n[k-1]}{\tau}$$

and we have

$$\frac{n[k] - n[k-1]}{\tau} = A\,n[k] + B\,p[k],$$

$$n[k] - \tau\,A\,n[k] = n[k-1] + \tau\,B\,p[k],$$

$$n[k] = [I - \tau A]^{-1}\,n[k-1] + [I - \tau A]^{-1}\,\tau\,B\,p[k],$$

$$n[k] = [I - \tau A]^{-1}\,[n[k-1] + \tau\,B\,p[k]], \qquad k = 1, 2, \ldots, N \to +\infty \quad (5.16)$$

Solving numerically the Eq. (5.16) we find approximate solution of the Eq. (5.6).

Asymptotic stability of the recurrence system (5.16) is a necassary condition of correctness of numerical calculations.

5.1.2.2 Asymptotic Stability of the Recurrence Scheme (5.16)

The system (5.16) is asymptotically stable if and only if all eigenvalues of the matrix $[I - \tau A]^{-1}$ have modules smaller than 1, i.e. $|\mu_i([I - \tau A]^{-1})| < 1$, $\forall i$, where μ_i denotes eigenvalues.

Let us consider the (triangular) matrix

$$I - \tau A = I - \tau \cdot \begin{bmatrix} -(\lambda + \frac{1}{s}) & 0 & 0 & 0 \\ \frac{1}{s} & \ddots & 0 & 0 \\ 0 & \ddots & \ddots & 0 \\ 0 & 0 & \frac{1}{s} & -(\lambda + \frac{1}{s}) \end{bmatrix} =$$

$$= \begin{bmatrix} 1 + \tau(\lambda + \frac{1}{s}) & 0 & 0 & 0 \\ -\frac{\tau}{s} & \ddots & 0 & 0 \\ 0 & \ddots & \ddots & 0 \\ 0 & 0 & -\frac{\tau}{s} & 1 + \tau(\lambda + \frac{1}{s}) \end{bmatrix}$$

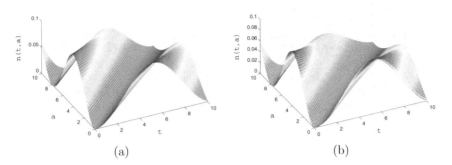

Fig. 5.3 Solutions of the McKendrick–Von Foerster equation for $\lambda = 0.1$ by the method of **a** characteristics **b** straight lines. Variables t and a discretization step the same in both cases: $\tau = s = 0.125$. Initial condition: $0.1 \cdot \sin^2(0.4 \cdot a)$, boundary condition: $0.1 \cdot \sin^2(0.3 \cdot t)$

The matrix $I - \tau A$ has one real multiple eigenvalue equals $1 + \tau(\lambda + \frac{1}{s})$. Therefore

$$\mu_i([I - \tau A]^{-1}) = \frac{1}{1 + \tau(\lambda + \frac{1}{s})}, \qquad i = 1, 2, \ldots, M$$

From the equation above one can see, that for $\tau > 0$, $s > 0$ i $\lambda > 0$ we have $\mu_i \in (0, 1)$. Consequently, the system (5.16) is asymptotically stable with any discretization steps $s > 0$ i $\tau > 0$.

Figure 5.3 presents examples of solutions of Eq. (5.6) obtained using the method of characteristics (Fig. 5.3a) and using the method of straight lines (Fig. 5.3b). It can be seen that as a solution we obtain the age-distribution function $n(t, a)$ evolving in time. It is now easy to interpret the formula (5.1) determining the total number of blood cells at time t. In Fig. 5.4, differences in results obtained using two methods can be compared.

5.2 Lasota–Ważewska Feedback

The originality of the Lasota–Ważewska model lies in the combination of the McKendrick–Von Foerster equation (5.6) with appropriately chosen feedback, with a delayed argument representing the $p(t)$ function of erythrocyte production. It is possible that this is the only equation of hematopoietic system dynamics in which the dependence on blood cell production was derived mathematically (relationships as such are very often determined on the basis of experimental data). We will now derive the production function found by Ważewska-Czyżewska and Lasota (1976, p. 26) (see also Rudnicki 2014, p. 219).

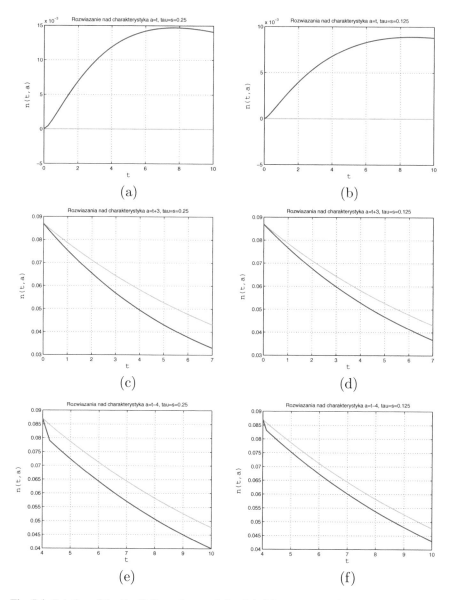

Fig. 5.4 Solution of the Eq. (5.6) on characteristics ((**a**) (**b**)): $a = t$ ((**c**) (**d**)): $a = t + 3$ ((**e**) (**f**)): $a = t - 4$. Green lines—method of characteristics, blue—method of straight lines. In figures **a**, **c**, **e**, variables t and a discretization step is 0.25, and the integral of the square error respectively 0.0014, 0.0004 and 0.0003. In figures (b), (d), (f) discretization step is 0.125, the integral of the square error respectively 0.001, 0.0003 and 0.0002

The increase in blood cell production per unit of time is described by the derivative $p'(t)$, therefore

$$S(t) = \frac{p'(t)}{p(t)} \tag{5.17}$$

denotes a unit increase of production, which we will call the degree of system stimulation. At equilibrium conditions $S(t) = 0$. The changing in the amount of erythrocytes in the bloodstream stimulates or stops their production (see Chap. 2). The system response (e.g. in the form of an increase in the number of red blood cells after their previous decrease) appears after the time needed to transform the undifferentiated bone marrow cell into a mature erythrocyte. M. Ważewska-Czyżewska and A. Lasota assumed (in order to obtain possibly simple model) that the degree of system stimulation is proportional to the change in the total number of blood cells in the bloodstream, i.e.

$$S(t) = -\gamma \frac{d}{dt} N(t - h), \tag{5.18}$$

where γ is a proportionality coefficient and h denotes a hematopoietic system delay of reaction (this is the time needed to develop mature erythrocyte—see remarks in Chap. 2.4). Thus, it can be seen that the loss of blood cells corresponds to the increase in the system stimulation and the increase in its inhibition. From formulas (5.17) and (5.18), we have

$$\frac{p'(t)}{p(t)} = -\gamma N'(t - h), \tag{5.19}$$

then after integration we receive

$$p(t) = \rho \cdot e^{-\gamma N(t-h)}, \tag{5.20}$$

where ρ is an intergation constant. Figure 5.5 presents the function $p(N_h) = \rho \cdot e^{-\gamma N_h}$, $N_h \equiv N(t - h)$ for parameters corresponding to the state of healthy human body, i.e. $\sigma = 0.01$ day^{-1}, $\gamma = 0.0015$ ml^{-1}, $\rho = 724.5$ ml/day (see Ważewska-Czyżewska and Lasota 1976, p. 32), (see Ważewska-Czyżewska 1983, p. 160,161). It can be seen that the function has a monotonically decreasing character. When the number of blood cells $N_h \equiv N(t - h)$ drops below the normal level $N = 2300$ ml, the intensity of blood cell production increases, after a delay time h, and when the number of blood cells returns to normal level, the production intensity decreases.

Putting together now (5.1), (5.6), (5.8) and (5.20) we obtain the Lasota–Ważewska equation describing dynamics of red blood cells system

$$\begin{cases} \dfrac{\partial n(t, a)}{\partial t} + \dfrac{\partial n(t, a)}{\partial a} = -\lambda(t, a) \cdot n(t, a), \\[2mm] n(t, 0) = p(t) = \rho \cdot e^{-\gamma \int_0^\infty n(t-h,\, a)\, da}. \end{cases} \tag{5.21}$$

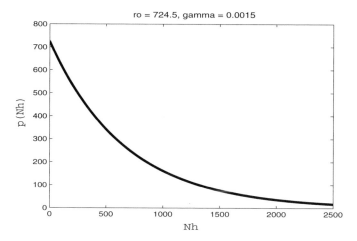

Fig. 5.5 Function $p(N_h) = \rho \cdot e^{-\gamma N_h}$, $N_h \equiv N(t-h)$ describing production of erythrocytes found by M. Ważewsk-Czyżewska and A. Lasota. Here, for parameters corresponding to the state of healthy human body, i.e. $\sigma = 0.01$ day^{-1}, $\gamma = 0.0015$ ml^{-1}, $\rho = 724.5$ ml/day

In the foregoing part, we provided an interpretation of the destruction coefficient λ, which denotes the probability that the blood cell, which at time t is at the age a will die in the time interval $(t, t + \Delta t)$. The meaning of the coefficient γ results from the formula (5.18). It determines the excitation degree $S(t)$ of the system caused by a unit change in the number of blood cells per unit of time. The ρ coefficient characterizes requirement of the body for an oxygen. The higher the demand, the higher the ρ. The h delay determines the time needed for the production of mature erythrocytes. The lengths of this time specified in literature are provided in Sect. 2.4.

5.3 Reduced Lasota–Ważewska Model

Partial differential equations modelling the dynamics of hematopoiesis can be reduced to the corresponding differential equations with delayed argument (see Ważewska-Czyżewska and Lasota 1976; Ważewska-Czyżewska 1983; Mackey and Milton 1990). As a result of the reduction, from the model taking into account the structure of e.g. the age of blood cells (i.e. such as the Lasota–Ważewska equation), we obtain a model describing the change in their total quantity (Ważewska-Czyżewska and Lasota 1976, p. 31) and (Rudnicki 2014, p. 219)). Mathematical information on differential equations with delayed argument can be found, among others, in books by Hale and Verduyn Lunel (1993) or Rudnicki (2014). Let us introduce the coefficient

$$\sigma = \frac{1}{N(t)} \int_0^\infty \lambda(t,a)n(t,a)da = \frac{\int_0^\infty \lambda(t,a)n(t,a)da}{\int_0^\infty n(t,a)da}. \qquad (5.22)$$

In numerator we have the number of blood cells destroyed per unit of time, in denominator, the total number of blood cells, therefore σ means a probability of blood cell destruction per unit of time. From the Eq. (5.22) results, that σ depends of t, however we will assume, that it is constant. Such simplification is made in order to formulate a model easier for mathematical analysis (see Rudnicki 2014, p. 205). Now integrating the Eq. (5.6), with respect to a in the interval $[0, \infty]$ we obtain

$$\int_0^\infty \frac{\partial}{\partial t} n(t,a)da + \int_0^\infty \frac{\partial}{\partial a} n(t,a)da = \int_0^\infty \lambda(t,a)n(t,a)da. \qquad (5.23)$$

Taking into consideration, that $N(t) = \int_0^\infty n(t,a)da$, then the first component can be written as $N'(t)$. Making natural biological assumption, that $\lim_{a \to \infty} n(t,a) = 0$, then the second component is $-n(t,0) = -p(t) = -\rho e^{-\gamma N(t-h)}$. From the Eq. (5.22) the third component can be replaced by the expression $-\sigma N(t)$. Therefore (5.23) takes the form

$$\frac{dN(t)}{dt} = -\sigma N(t) + \rho e^{-\gamma N(t-h)}. \qquad (5.24)$$

It is the so-called the reduced Lasota–Ważewska equation. We have already given the interpretation of all constants present in this equation.

Equation (5.24) can be written in the general form (see Ważewska-Czyżewska 1983, p. 157 and further)

$$\frac{dN(t)}{dt} = -D(t) + P(t), \qquad (5.25)$$

where $D(t)$ denotes level of blood cells destruction at time t, and $P(t)$ level of their production at time t. In the considered case $D(t) = \sigma N(t)$, $P(t) = \rho e^{-\gamma N(t-h)}$.

5.3.1   Mackey–Glass Equation

Mackey and Glass (1977), in their model of changes in the amount of circulating red blood cells in the bloodstream, proposed a bit different form of production function

$$\frac{dN(t)}{dt} = -\sigma N(t) + \frac{\beta_0 \theta^n}{\theta^n + N_h^n}, \quad N_h \equiv N(t-h). \qquad (5.26)$$

It has however the same monotonically decreasing character as the function obtained by M. Ważewska-Czyżewska and A. Lasota (compare Figs. 5.5 and 5.6). Information how it was formulated can be found in an article by Mackey (1978) and in a book

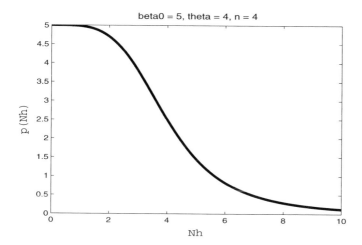

Fig. 5.6 Nonlinearity determining the production function in the Mackey–Glass equation (5.26)

by Ważewska-Czyżewska (1983, p. 159). The biological interpretation of the β_0 constant from the Eq. (5.26) corresponds to the interpretation of the constant ρ from the Eq. (5.24), and the interpretation of θ corresponds to the interpretation of γ.

The dynamics of the hematopoietic system have been considered by many authors. There is a lot of models in the literature distinct in terms of mathematical construction and bio-medical conditions taken into consideration. E.g. papers of (Ackleh et al. 2006; Adimy and Crauste 2009, 2003; Adimy et al. 2005a, c; Bernard et al. 2003) concern structural models, while e.g. works of (Adimy et al. 2006a, 2005b, 2006b; Crauste 2006; Adimy et al. 2008) concern models in a form of delay equations. The content presented in this book in Chaps. 2 and 5 has been presented in detail so that the reader can find significant differences between existing models of hematopoiesis and the model considered in this monograph in the next Chap. 6.

Chapter 6
Lasota Equation with Unimodal Regulation

Clinical studies on hematopoiesis show that production level $P(t)$ of blood cell elements in Eq. (5.25) can have different than monotonically decreasing character (see e.g., Ważewska-Czyżewska 1983; Mackey and Milton 1990). In some cases, it may demonstrate non-monotonic character, or in other words it may be described by a function with one smooth maximum for arguments greater than zero. Such functions are called unimodal (see e.g., Röst and Wu 2007; Liz and Röst 2009). Let us define what we will mean by the unimodal function.

Definition 6.22 Let us consider the following conditions:

$$\forall x \geq 0 \ f(x) \geq 0, \ f(0) = 0 \text{ and}$$
$$\exists x_0 > 0 \text{ such, that } f'(x_0) = 0 \text{ and} \qquad (6.1)$$
$$f'(x) > 0 \text{ for } 0 \leq x < x_0 \text{ and } f'(x) < 0 \text{ for } x > x_0.$$

$$\lim_{x \to \infty} f(x) = 0. \qquad (6.2)$$

$$f''(x) < 0 \text{ for } 0 \leq x \leq x_0. \qquad (6.3)$$

$$\exists x_p > 0 \text{ such, that } f''(x) > 0 \text{ for } 0 < x < x_p \text{ and}$$
$$f''(x) < 0 \text{ for } x_p < x < x_0. \qquad (6.4)$$

We will call function $f : [0, \infty) \to [0, \infty)$ unimodal if it fulfills conditions (6.1) and (6.2) and additionally condition (6.3) or (6.4).

Figure 6.1a shows function fulfilling conditions (6.1) and (6.2) and (6.3), and Fig. 6.1b fulfilling conditions (6.1) and (6.2) and (6.4).

© The Editor(s) (if applicable) and The Author(s), under exclusive license to Springer Nature Switzerland AG 2021
P. J. Mitkowski, *Mathematical Structures of Ergodicity and Chaos in Population Dynamics*, Studies in Systems, Decision and Control 312, https://doi.org/10.1007/978-3-030-57678-3_6

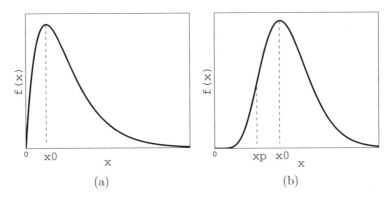

Fig. 6.1 Unimodal functions **a**: function fulfilling conditions (6.1) and (6.2) and (6.3) from the Definition 6.22 **b**: function fulfilling conditions (6.1) and (6.2) and (6.4) from the Definition 6.22

6.1 Lasota Equation for Erythrocytes

Ważewska-Czyżewska (1983, p. 165) stated, that unimodal dependance $P(t)$ from the quantity of circulating in the bloodstream erythrocytes is very unlikely, but it can occur in several acute pathological cases or e.g., when the body is near death. Lasota (1977) formulated a model with unimodal feedback with a delayed argument to understand the origin of irregular changes in the amount of erythrocytes circulating in the bloodstream. The equation has the form:

$$\frac{dN(t)}{dt} = -\sigma \cdot N(t) + (\rho \cdot N(t-h))^s \cdot e^{-\gamma \cdot N(t-h)} \tag{6.5}$$

and is the main subject of research in this book. The production of erythrocytes is given by the feedback in the form of a unimodal function $p(N_h) = (\rho \cdot N_h)^s \cdot e^{-\gamma \cdot N_h}$, $N_h \equiv N(t-h)$ with delayed argument. This function for $0 < s \leq 1$ fulfills conditions (6.1) and (6.2) and (6.3) from the Definition 6.22 (see Fig. 6.2a), while for $s > 1$ fulfills conditions (6.1) and (6.2) and (6.4) from the Definition 6.22 (see Fig. 6.2b).

In these cases, the function has an extremum (maximum) at point A with the abscissa $x = -\frac{s}{\gamma}$, and inflection points C and D with the abscissa $x = -\frac{s \pm \sqrt{s}}{-\gamma}$ (see Bronsztejn et al. 2004). The bio-medical interpretation of constants σ, ρ, γ and h is the same as for the Lasota-Ważewska equation (5.24). The interpretation of the power s was given by the author of this book in (Mitkowski 2011), (see also Mitkowski 2012). Let us recall it here. We will use the inverse reasoning to this used by M. Ważewska-Czyżewska and A. Lasota, in which from the so-called system stimulation level, they derived a feedback determining the production of erythrocytes (see Sect. 5.2). This time we already have formula for blood cells production and we will look for a corresponding system stimulation level in the form

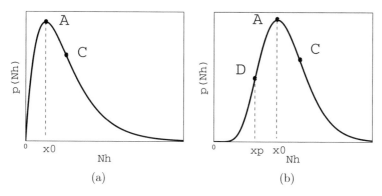

Fig. 6.2 Unimodal feedback of the Lasota equation (6.5) **a**: when $0 < s \leq 1$ **b**: when $s > 1$

$$S(t) = \frac{p'(t)}{p(t)}.$$

One must therefore find $p'(t)$ for

$$p(t) = (\rho \cdot N(t-h))^s \cdot e^{-\gamma \cdot N(t-h)}. \tag{6.6}$$

Let us write

$$p(t) = p_1(t) \cdot p_2(t), \tag{6.7}$$

where

$$p_1(t) = (\rho \cdot N(t-h))^s, \quad p_2(t) = e^{-\gamma \cdot N(t-h)}. \tag{6.8}$$

We have therefore

$$p'(t) = p_1'(t) \cdot p_2(t) + p_1(t) \cdot p_2'(t), \tag{6.9}$$

where

$$p_1'(t) \cdot p_2(t) = \rho^s s (N(t-h))^{s-1} N'(t-h) e^{-\gamma N(t-h)}, \tag{6.10}$$

$$p_1(t) \cdot p_2'(t) = -\gamma e^{-\gamma N(t-h)} N'(t-h)(\rho N(t-h))^s. \tag{6.11}$$

I.e.

$$S(t) = \frac{p'(t)}{p(t)} = \frac{(\rho N(t-h))^s e^{-\gamma N(t-h)} [-\gamma + s \frac{1}{N(t-h)}]}{(\rho N(t-h))^s e^{-\gamma N(t-h)}} N'(t-h), \tag{6.12}$$

and finally we have

$$S(t) = \frac{p'(t)}{p(t)} = -\gamma N'(t-h) + s \frac{N'(t-h)}{N(t-h)}. \tag{6.13}$$

When $s = 0$ then we have earlier considered Eq. (5.24), where erythrocytes production function has a monotonically decreasing character, physiologically natural, for the normal erythropoiesis. When $s > 0$ then the production function is distorted to the unimodal function, and the segment $-\gamma N'(t - h)$ of the Eq. (6.13) responsible for the normal erythropoiesis is inhibited by the segment $s \frac{N'(t-h)}{N(t-h)}$, where the expression

$$Z(t) = \frac{N'(t - h)}{N(t - h)} \tag{6.14}$$

denotes a relative change in the number of blood cells. Thus s represents the degree of disorder of the normal erythropoietic response. When $s = 0$ then the response is normal, when $s > 0$, then the response is inhibited and the greater is the s, the greater is the inhibition.

6.1.1 Mackey-Glass Equation for Neutrophils

Mackey and Milton (1990, p. 9) report that the dependence of the level of neutrophil production from their amount in the bloodstream is also unimodal. Neutrophils are one of the type of white blood cells (Craig et al. 2010). The model for neutrophils was formulated by Mackey and Glass (1977) and has the form:

$$\frac{dN(t)}{dt} = -\sigma \cdot N(t) + \frac{\beta_0 \theta^n N_h}{\theta^n + N_h^n}, \quad N_h \equiv N(t - h) \tag{6.15}$$

Unimodal feedback (see Fig. 6.3) fulfills conditions (6.1) and (6.2) and (6.3) from the Definition 6.22.

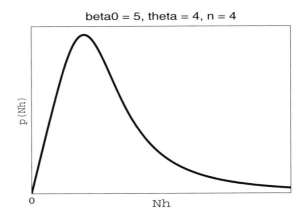

Fig. 6.3 Unimodal feedback in the Mackey'a-Glass'a equation (6.15) for netrophils

beta0 = 5, theta = 4, n = 4

p (Nh)

0   Nh

6.1.2 Nicholson Australian Sheep Blowfly Equation

Gurney et al. (1980) proposed a model to analyze the experimental data described by Nicholson (1954) concerning the population of the Australian sheep blowfly (*lucilla cuprima*) (see also Lasota 1977, p. 243 and Murray 2006, p. 18). Model has a form of Lasota equation (6.5) with $s = 1$, i.e.

$$\frac{dN(t)}{dt} = -\sigma \cdot N(t) + (\rho \cdot N(t - h)) \cdot e^{-\gamma \cdot N(t-h)}, \qquad (6.16)$$

therefore, the unimodal nonlinearity satisfies conditions (6.1) and (6.2) and (6.3) form the Definition 6.22. Equation (6.16) has been studied by many researchers in various approaches (see e.g., So 1998; Yi and Zou 2008; Li et al. 2007; Zhang and Peng 2008; Alzabut 2010; Wang and Wei 2008; Bradul and Shaikhet 2007).

Buedo-Fernández and Liz (2018) emphasized the fact that the Eq. (6.5) with different values of s was applied to model various real phenomena. For $0 < s < 1$ Matsumoto and Szidarovszky (2013) used it as an extension of Solow (1956) and Swan (1956) neoclassical model of economic growth. In turn Huang at al. (2014) and Liz and Ruiz-Herrera (2015) applied Eq. (6.5) with $s > 1$ to model dynamics of single-species populations exhibiting the Allee effect. Following the thought of A. Lasota, which is placed at the very beginning of this book as its motto, seems that the dynamical structure of the Lasota equation (6.5) could be the part of the reality structure itself.

The Lasota equation (6.5) is structurally (mathematically) distinguished from the biological models (6.15) and (6.16) by the fact that the unimodal nonlinearity used in it at $s > 1$ can meet the conditions (6.1) and (6.2) and (6.4) from the Definition 6.22, i.e. it can have an inflection point D as in Fig. 6.2b, which is not possible in the case of nonlinearity from the Mackey-Glass equation for neutrophils (6.15) or the Gurney equation et al. Eq. (6.16) for the Nicholson's Australian sheep blowfly population (i.e. Lasota equation (6.5) with $s = 1$). In addition, there is a difference in bio-medical conditions. The Lasota equation applies to red blood cells, the Mackey-Glass equation to neutrophils (white blood cells), and the Gurney et al. equation to Australian sheep blowfly, so the formulated bio-medical interpretations must be different. In the Lasota equation (6.5) with $s > 1$, we have two asymptotically stable fixed points (see Fig. 6.4) $N1 = 0$ and $N3 > 0$ and the unstable point $N2$ ($N1 < N2 < N3$). The bio-medical interpretation of this fact is the following (Lasota 1977, p. 245). Under normal conditions, the number of red blood cells remains constant (fixed point $N3$). In diseases such as e.g., periodic hematopoiesis, the number of blood cells oscillates periodically (see Mackey and Milton 1990, p. 17) around this constant level, and in pathological situations it may oscillate non-periodically. However, when the amount of blood cells drops below a certain level, e.g., as a result of some rapid bleeding, then the body dies, what corresponds to the attraction of trajectories, starting "below" a certain value, to the point $N1 = 0$. Numerical simulations indicate that for low values of delay h, this value is determined by the position of the unstable point $N2$ (as an example see Figs. 6.5b and 6.6a, b). For higher values of h and constant initial

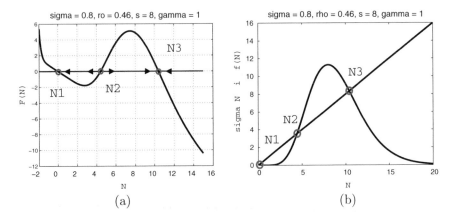

(a)   (b)

Fig. 6.4 Fixed points for the Eq. (6.5) when $s > 1$. Points $N1$ and $N3$ are asymptotically stable, and point $N3$ is unstable. **a**: Right-hand side of the equation **b**: Linear part σN and unimodal nonlinearity

functions, the $N2$ point still determines this value, whereas for initial functions of a different character, it is not exactly at this point but lies near it. The construction of the Lasota equation (6.5) provides one more interesting bio-medical interpretation. When $s = 0$, we get the reduced Lasota-Ważewska model (5.24) with the entire theory of periodic solutions of this equation, while when $s > 0$, the equation besides periodic allows also chaotic solutions.

6.2 Lasota Hypothesis

Lasota (1977, p. 248) formulated the hypothesis about the existence of nontrivial ergodic properties for the Eq. (6.5). Let C_h be a space of continuous functions $v: [-h, 0] \to R$ with the supremum norm topology. For some positive values of the parameters ρ, h, s and σ there exists on C_h continuous measure which is ergodic and invariant with respect to Eq. (6.5). Measure μ is called continuous if it vanishes on points and in this sense, it is nontrivial.

The hypothesis in the language of ergodic theory raises the question about the existence of chaotic solutions of the differential Eq. (6.5) with delay. It also looks like a generalization of Ulam (1960) hypothesis concerning nontrivial ergodic properties of the section [0, 1] into itself map (see Sect. 4.3) on differential equations with delay of type $\dot{x}(t) = f(x(t), x(t - h))$. Such coupling appears during computational analysis of the Eq. (6.5) properties, where in search of ergodic features its right-hand side "shape" is suitably chosen. The Ulam hypothesis contains the question of what "shape" must have a function that defines the section [0, 1] into itself map, in order for this map to have non-trivial ergodic properties.

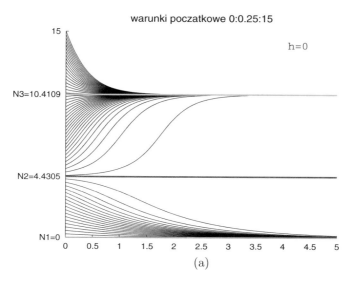

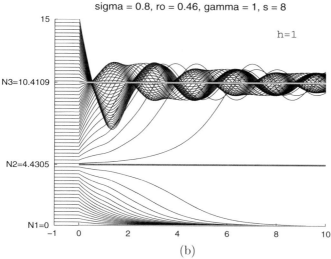

Fig. 6.5 Trajectories of the Eq. (6.5) $\sigma = 0.8$, $\rho = 0.46$, $s = 8$, $\gamma = 1$ **a**: when the delay $h = 0$ **b**: when $h = 1$ and initial functions have constant values on $[-h, 0]$

6.3 Computational Verification of Lasota Hypothesis

Computational studies (Mitkowski 2011, 2012; Mitkowski and Mitkowski 2012) suggest that the Eq. (6.5) exhibits nontrivial ergodic properties for

$$\rho \in [0.46, \ 0.52], \ \sigma = 0.8, \ s = 8, \ \gamma = 1. \qquad (6.17)$$

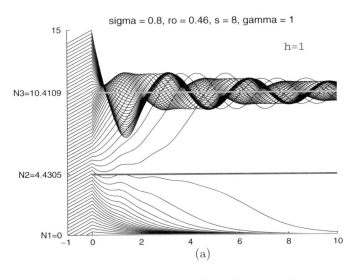

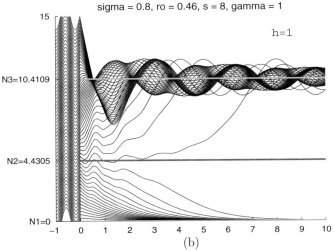

Fig. 6.6 Trajectories of the Eq. (6.5) $\sigma = 0.8$, $\rho = 0.46$, $s = 8$, $\gamma = 1$, $h = 1$ **a**: $h = 1$ and initial functions of type $a \cdot t + b$ **b**: $h = 1$ and initial functions of type $a \cdot \sin(b \cdot t) + c$

In addition, the delay h must be sufficiently large, i.e. $h > 9$. Figures 6.7, 6.8, 6.9 show changing character of the Eq. (6.5) solution projected on the plane $N(t) \times N(t - h)$ with increasing value of delay h. When $h \approx 10$ and larger, the trajectory appears to fill densely some area of the space, what may already be a preliminary premise for the process ergodicity (see Chap. 4 and e.g., (Kudrewicz 1991, p. 189)). The nature of the solution changes can also be shown in the graph of the number of trajectory intersections with a properly selected one-dimensional axis. In Figs. 6.7, 6.8 and 6.9, this axis is marked with a blue line and passes through the stationary point $N3$. The trajectory intersection points with this axis, as the delay h increases, are shown in Fig. 6.10. Simulations indicate that for nontrivial ergodic properties, the delay h can take large values, for example, $h = 50$, but the greater its value is, the greater is the trajectory attraction to the point $N1 = 0$, and for very large h, nontrivial ergodic properties disappear (see Figs. 6.11, 6.12, 6.13, 6.14, and 6.15). Figure 6.16a shows the range of the right-hand side of the Eq. (6.5) for the given parameters in (6.17). If we additionally draw there a straight line $y = x$, as in the Ulam hypothesis (see Sect. 4.3 and Fig. 4.18), it seems that structurally problems from Lasota hypothesis and Ulam hypothesis become similar. In the case of Ulam they relate to the section $[0, 1]$ into itself map, and in the Lasota hypothesis, the differential equation with delay. In this sense, we said earlier that the Lasota hypothesis is a generalization of the Ulam hypothesis for equations with delay of type $\dot{x}(t) = f(x(t), x(t - h))$. Figure 6.16b shows the range of unimodal nonlinearity of the Eq. (6.5) with the linear component σN drawn in.

6.3.1 Introduction to Computational Analysis

To study the ergodic properties of any system by means of numerical experiments, a large set (flow) of its trajectories must be generated, just as we did for the logistics mapping in Sect. 4.1.2. With a large set of trajectories, it is possible to study its average behavior and compare it with the average along individual trajectories (see Sect. 4.1).

We will now present selected simulation results of the Eq. (6.5) for

$$\rho = 0.46, \ \sigma = 0.8, \ s = 8, \ \gamma = 1, \ h = 10, \qquad (6.18)$$

this is the lower bound of the hatched areas in Fig. 6.16a, b. All trajectories of the Eq. (6.5) were calculated (with the DDE23 solver), as well as all other numerical computations and illustrations, were mace using MATLAB software. For the remaining values of ρ from the interval in (6.17) and for fairly large range of $h > 9$, the obtained numerical results qualitatively have the same character, while there are some quantitative differences.

Figure 6.17 presents three stages of the evolution of the flow of large number of Eq. (6.5) trajectories. They start from initial functions with constant, close values, uniformly distributed over a certain subinterval R^+. One can notice that the flow is first quite regular, after some time t gets complicated, and finally becomes very irregular, with trajectories being clearly limited. First, we will try to numerically approximate the evolution of the density of the distribution in various spaces and check if it will converge to some smooth density that does not change for longer simulation times. We have done the same thing in Sect. 4.1.2 for the logistic mapping (4.3). However, for the logistic mapping, the exact density evolution can be determined by calculating the subsequent iterations of the Frobenius–Perron operator (see Sect. 4.1.2). It is also possible to determine precisely the limiting density. In the case

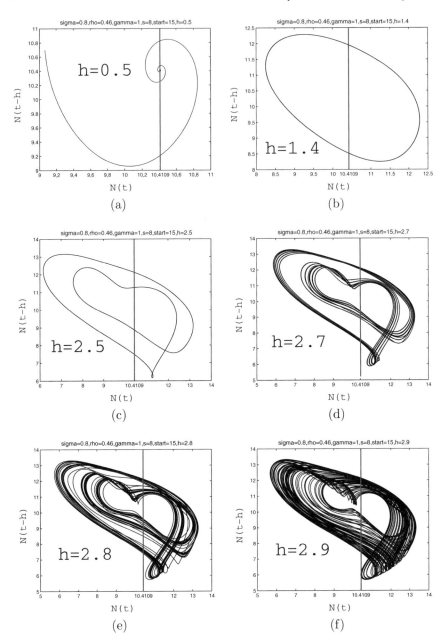

Fig. 6.7 Bifurcations of the Eq. (6.5) solutions projected on a plane $N(t) \times N(t - h)$ for parameters $\sigma = 0.8$, $\rho = 0.46$, $s = 8$, $\gamma = 1$ and initial function with a constant value of 15, with an increase of delay h **a**: $h = 0.5$ **b**: $h = 1.4$ **c**: $h = 2.5$ **d**: $h = 2.7$ **e**: $h = 2.8$ **f**: $h = 2.9$

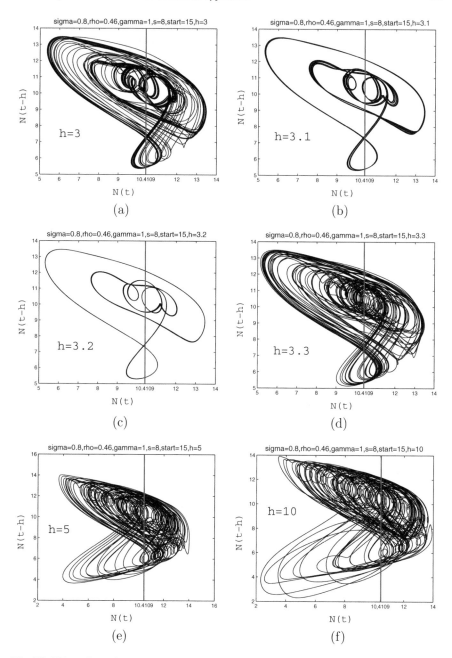

Fig. 6.8 Bifurcations of the Eq. (6.5) solutions projected on a plane $N(t) \times N(t - h)$ for parameters $\sigma = 0.8$, $\rho = 0.46$, $s = 8$, $\gamma = 1$ and initial function with a constant value of 15, with an increase of delay h **a**: $h = 3$ **b**: $h = 3.1$ **c**: $h = 3.2$ **d**: $h = 3.3$ **e**: $h = 5$ **f**: $h = 10$

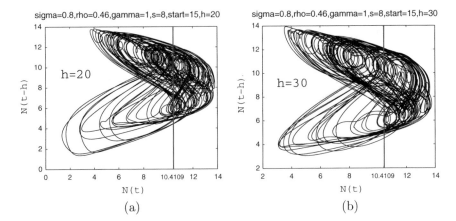

Fig. 6.9 Solution of the Eq. (6.5) projected on a plane $N(t) \times N(t - h)$ for parameters $\sigma = 0.8$, $\rho = 0.46$, $s = 8$, $\gamma = 1$ and initial function with a constant value of 15 for the delay **a**: $h = 20$ **b**: $h = 30$

of Eq. (6.5), it is not very clear whether it is possible to formulate the Frobenius–Perron operator at all, which is why we approximate numerically the evolution of density. One more problem here is worth paying attention. Logistic mapping (4.3) has the so-called absolutely continuous invariant measure determined by Ulam and von Neumann (1947) (see Sect. 4.1.2) with the density given by the formula (4.7). Lasota and Yorke (1977) also showed that the logistic mapping for $\alpha < 4$ has the so-called continuous (vanishing on points) invariant and ergodic measure (see also Lasota 1977, 1978; Myjak 2008). It is known that an absolutely continuous measure can be obtained by numerical approximation of the Frobenius–Perron operator (see e.g., Ding and Zhou 1996). However, we do not have a good idea of how the "continuous measure" could look like numerically. The problem of computational approximation of the Frobenius–Perron operator was raised by Ulam (1960) and then analyzed for various classes of systems by among others Li (1976), Ding et al. (1993), Ding and Zhou (1996; 1999), Mohseni-Moghadam'a and Panahi (2000), Ding et al. (2002), Panahi (2007). In particular Taylor (2004) considered the approximation of the Frobenius–Perron operator for delayed differential equations.

In this book, we will approximate the evolution of density by using the simplest way, i.e. by calculating a very large set of Eq. (6.5) trajectories. Useful remarks regarding the numerical determination of density evolution can be found in books of Lasota and Mackey (1994), Kudrewicz (1991,1993, 2007) and Ott (1997), as well as in the paper (Mitkowski and Ogorzałek 2010).

The monograph of Losson et al. (2020) concerns examination of the densities evolution in systems whose dynamics are described by differential delay equations.

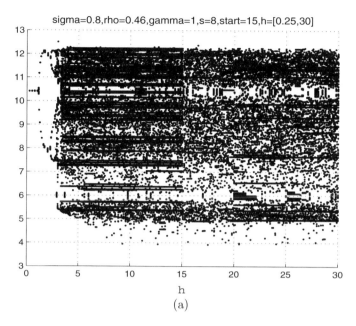

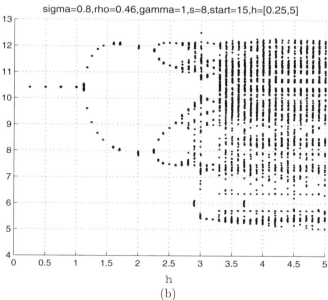

Fig. 6.10 **a**: Equation (6.5) trajectory intersection diagram projected on a plane $N(t) \times N(t - h)$ with one-dimensional axis passing through the stationary point $N3$ (blue axis in Figs. 6.7, 6.8, 6.9), at $h \in [0, 30]$ with a discretization step 0, 1. **b**: The same diagram for $h \in [0, 5]$. Trajectories were calculated for $\sigma = 0.8$, $\rho = 0.46$, $s = 8$, $\gamma = 1$ and initial function with constant value of 15

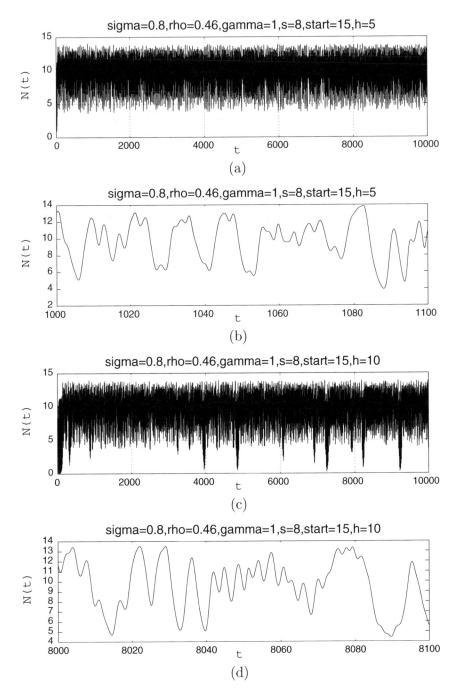

Fig. 6.11 With increasing delay h the trajectory is more and more attracted to the stationary point $N1 = 0$ (this is more clearly visible at higher values of h in Figs. 6.12, 6.13, 6.14 and 6.15). Here **a**: trajectory for $h = 5$ **b**: its time course in a shorter time interval **c**: trajectory for $h = 10$ **d**: its time course in a shorter time interval

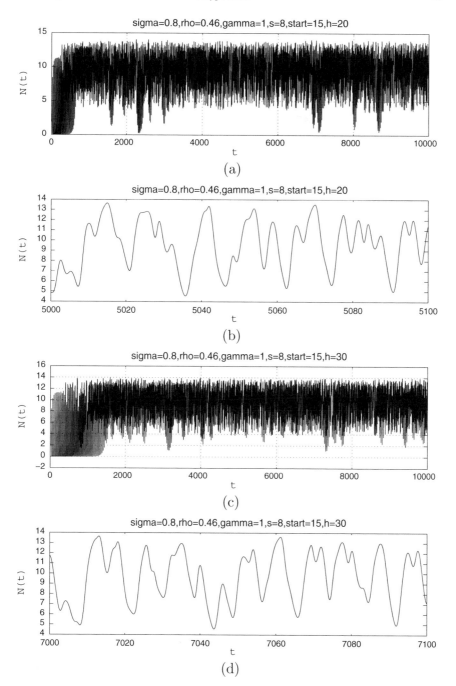

Fig. 6.12 With increasing delay h the trajectory is more and more attracted to the stationary point $N1 = 0$. **a**: Trajectory for $h = 20$ **b**: its time course in a shorter time interval **c**: trajectory for $h = 30$ **d**: its time course in a shorter time interval

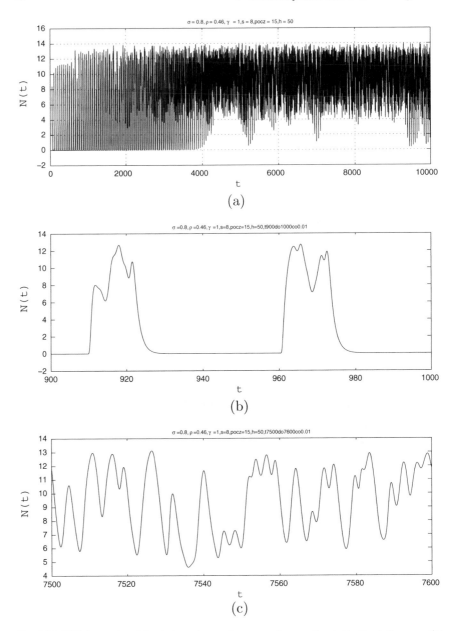

Fig. 6.13 With increasing delay h the trajectory is more and more attracted to the stationary point $N1 = 0$. **a**: Trajectory for $h = 50$ **b**: its time course in the time interval [900, 1000] **c**: its time course in the time interval [7500, 7600]

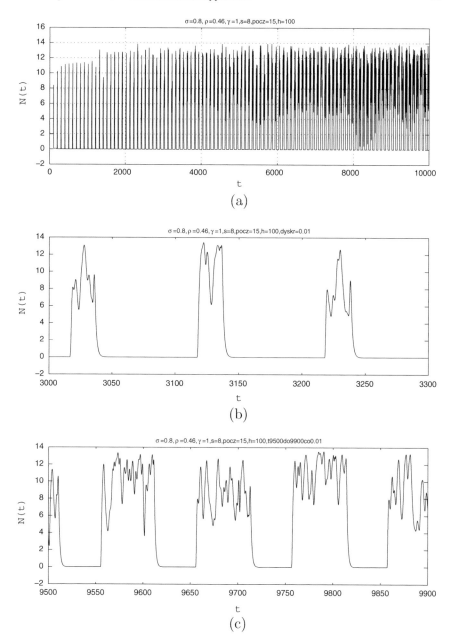

Fig. 6.14 With increasing delay h the trajectory is more and more attracted to the stationary point $N1 = 0$. **a**: Trajectory for $h = 100$ **b**: its time course in the time interval [3000, 3300] **c**: its time course in the time interval [9500, 9900]

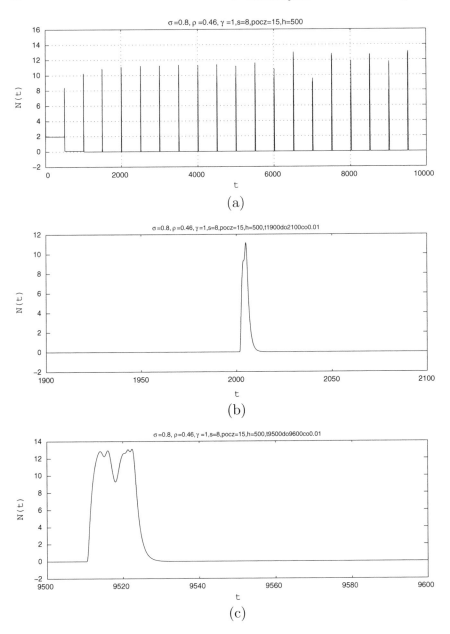

Fig. 6.15 With increasing delay h the trajectory is more and more attracted to the stationary point $N1 = 0$. **a**: Trajectory for $h = 500$ **b**: its time course in the time interval $[1900, 2100]$ **c**: its time course in the time interval $[9500, 9600]$

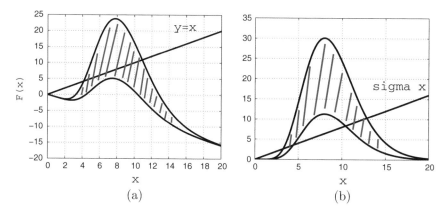

Fig. 6.16 Range of parameters (6.17) graphically presented (area covered with blue diagonal lines) (**a**): right hand-side of the Eq. (6.5), i.e. $F(x) = -\sigma \cdot x + (\rho \cdot x) \cdot e^{-\gamma \cdot x}$ (**b**): linear component $\sigma \cdot x$ and unimodal function $f(x) = (\rho \cdot x) \cdot e^{-\gamma \cdot x}$

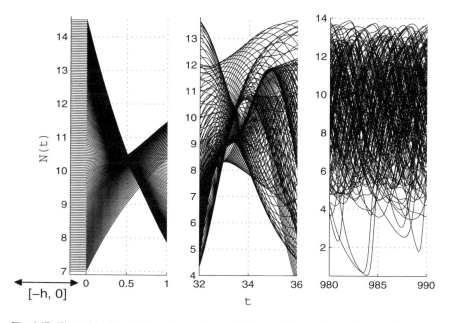

Fig. 6.17 Flow of the Eq. (6.5) trajectories for $\rho = 0.46$, $\sigma = 0.8$, $s = 8$, $\gamma = 1$, $h = 10$, starting from constant initial functions with values distributed uniformly on a subinterval R^+. First the flow is regular, but after some time it becomes turbulent

6.3.2   Defining the Space for Analysis

The solution of the Eq. (6.5) can be analyzed in various ways. The change in the instantaneous value of the trajectory can be examined, but you can also treat the solution of this equation as a "time-shifting" state in the form of a function specified on the delay h length interval, and therefore analyze the change of the entire state. The selection depends on the purpose of the research. From the application point of view, the analysis of instantaneous values would refer to measuring the changing quantity at a given time. However, no physical value can in practice be measured in an infinitely short time. We rather suspect that the measurement result may refer to some time interval, maybe even a very short one, but still an interval. Therefore, when we take a delay h length state (function) and specify it in some norm, e.g., L^1 (i.e. we calculate the integral from the state), we get a kind of "averaged" value of the state and thus representation of the solution on a certain time interval. In this sense, the selection of the norm, and more generally the space of solutions may depend on the purpose of the research.

In the computational studies of the Eq. (6.5), we will consider selected subspaces of infinitely dimensional space of all equation state points. The subspace of this space is the space of instantaneous values $N(t) \in R$. Figure 6.18a shows an example state of the equation, and Fig. 6.18b its geometric representation in the example six-dimensional subspace of the infinitely dimensional space of all state points. In such representation, the function becomes a point, and its evolution described by Eq. (6.5) will be illustrated by the movement of this point in space.

Looking at Fig. 6.18, a geometric representation of state evolution in infinitely dimensional space can be imagined. In the following numerical calculations, we will deal with the analysis of dynamics in the spaces $N(t) \in R$, $N(t) \times N(t - h/2)$ and $N(t) \times N(t - h/2) \times N(t - h)$ because in one, two and three dimensions, one can quite clearly illustrate changes in the system behavior. The space $N(t) \times N(t - h)$ is very often used for imaging the delay differential equations dynamics (see e.g., Murray 2006; Bodnar and Foryś 2009).

Systems that exhibit ergodic properties (e.g., mixing property) in the entire considered space also exhibit these properties in subspaces. For such systems, convergence to a statistically steady state in subspaces is faster than in the entire space (see Dorfman 2001). We observed this phenomenon on simple low-dimensional mixing mappings in the Sect. 4.2.2. Therefore, if the Eq. (6.5) showed in the entire space characteristic properties of mixing systems, then it should also display such properties in selected subspaces.

In numerical studies we will present the state of the Eq. (6.5) also in supremum, L^1 and L^2 norms.

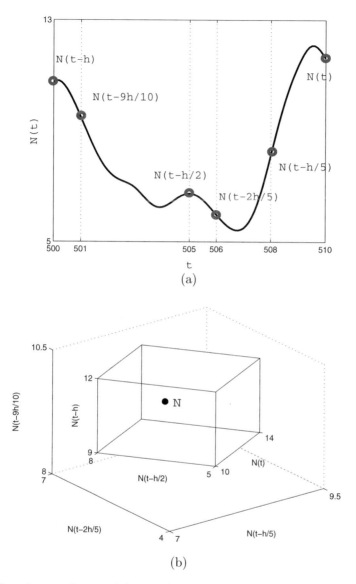

Fig. 6.18 a: An example state of the Eq. (6.5) **b**: its geometric representation in the example six-dimensional subspace $N(t) \times N(t - h/2) \times N(t - h) \times N(t - h/5) \times N(t - 2h/5) \times N(t - 9h/10)$ of the infinitely dimensional space of all state points. It can be seen, that in this space the equation state is a point

6.3.3   Ergodic Properties of the System

Initially we examine whether system (6.5) shows the most characteristic property of an ergodic system, i.e. whether the average value over a large set of trajectories is equal to the average along all individual trajectories. Formally, according to Birkhoff theorem (see Chap. 4), the average value over a set is to be equal to the average along almost all trajectories, but in numerical studies, the probability that we will choose a trajectory that does not meet the ergodicity condition (of course, when the considered system is ergodic) is extremely small, in some references one can find information that this probability is 0 (see Kudrewicz 1991; Ott 1997). In the first place, we examine the behavior of approximated (from a set of 10^4 trajectories) initial distribution densities (we analyze various densities). We check whether the initial densities converge to smooth density that does not change in time, approximating potentially existing limiting density. Figures 6.19, 6.20, 6.21 show simulation stages (at selected moments in time) of the evolution of the constant initial function exponential distribution (in other words, the selected stages of the simulation of the evolution of the exponential initial distribution) on the subinterval R^+. In the left column of the figures (from top to bottom: Fig. 6.19a, c, e then Fig. 6.20 a, c, e, and further Fig. 6.21a, c), we can see the evolution on the space of instantaneous values $N(t) \in R$. In the right column (also from top to bottom Fig. 6.19b, d, f, then Fig. 6.20b, d, f, and further Fig. 6.21b, d) the evolution on the space $N(t) \times N(t - h)$ is shown. One can notice that after some time approximated densities seem actually not to changes their "shape". Densities are approximated by determining histograms that count the number of values from the set of trajectories in a given subinterval of the considered space and at a given time, which are then normalized to histograms with a surface area (in the case of histograms counting values on $N(t) \in R$) or volume (in the case of histograms on $N(t) \times N(t - h)$) equal to 1. Other distributions of constant initial functions were also studied and the same invariant "limiting" histograms were always obtained. Therefore, we obtain invariant approximated densities, however it is difficult to state from the shown figures whether they are smooth. To examine this more accurately with the method we used, it would be necessary to significantly increase the number of trajectories in the examined set, what can make numerical calculations time consuming. However, if the approximated means along single trajectories were similar to obtained limiting densities for the set of trajectories, this would indicate the ergodic properties of the system. Then, we could calculate very long trajectories, what is much less time-consuming, and verify if the means are smoothing out. By doing so, assuming the system ergodicity, we could also obtain an approximation of the average over the set, and it would also be a very nice example of the utility of the system ergodic properties. And indeed for single trajectories, the approximated average time that they spend in the subsets of the considered spaces is very similar to the average over large sets of trajectories (see Fig. 6.22). Histograms of averages are approximated according to the same principle as for a set of trajectories, but now the number of values along single long trajectory is counted. In Fig. 6.22a, b, approximations of the average for 10^4 points of trajectory are presented, and in Fig. 6.22c, d for 10^6

points. We can see that the histograms are smoothing out. Figure 6.22e shows a "from above" view in a grey scale of the histogram from the Fig. 6.22d. Demonstrating it in such way more clearly can be seen that the trajectory appears to densely fill a certain area of the space. We can also see (Fig. 6.22e) that the time course of the trajectory from which histograms were calculated is very irregular. The results presented here were obtained for the trajectory starting from the initial function with a constant value of 8. Trajectories for different types of initial functions were examined, e.g., $a \cdot x + b$, $a \cdot \sin(b \cdot x) + c$ or pseudo-random waveforms obtained using RAND or RANDN MATLAB software functions. In all cases, the results were the same as presented here for the initial function with a constant value, what is characteristic for ergodic processes (Fig. 6.23).

6.3.3.1 Distinguishing Between Chaotic and Periodic Solutions in Numerical Experiments

The following important fact should be noted. To identify behaviour that may in computational studies indicate non-trivial ergodic properties of the system, results should be observed from different perspectives, e.g., analysis of instantaneous values $N(t)$ histograms only is not sufficient. Let us consider the periodic solution of the Eq. (6.5), which in numerical simulations, can be obtained for parameters provided in (6.18), but with a delay, e.g., $h = 2$. The time course of the trajectory corresponding to these parameters (Fig. 6.24a), its representation on the plane $N(t) \times N(t - h)$ (Fig. 6.24c) and the histogram on this plane (Fig. 6.24d) indicate, that the trajectory is periodic. If, however, we did not look at these figures and only analyzed the histogram counting the number of instantaneous values $N(t)$ (Fig. 6.24b), then it could suggest that the trajectory densely fills a certain area of space. Additionally this histogram resembles a histogram for a chaotic trajectory of the logistic mapping (4.3) (Fig. 4.4) approximating an absolutely continuous invariant measure. It is, therefore, necessary to observe simulation results comprehensively and from different perspectives. In case of histograms presented earlier, time courses of the trajectories are very irregular. Exemplary waveform is presented in the Fig. 6.22e.

Thus, numerical simulations indicate that on the subspaces $N(t)$, $N(t - h)$ and $N(t) \times N(t - h/2) \times N(t - h)$ of the infinitely dimensional space of all the system (6.5) state points, the system display non-trivial (not focused on a point) ergodic properties. It is also true, when we consider the supremum norms, L^1 and L^2 on the state. The left column of the Fig. 6.25 contains numerically determined histograms approximating limiting densities for sets of trajectories calculated as a sequence of norms from an evolving state (Fig. 6.25 (a) supremum norm (b) L^1 norm (c) L^2 norm). The right column of Fig. 6.25 presents histograms for individual trajectories.

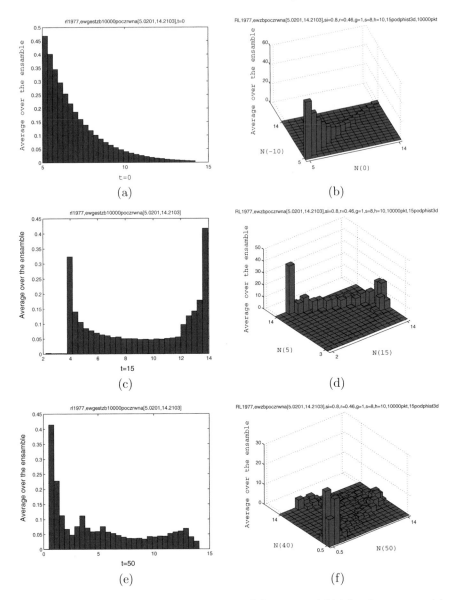

Fig. 6.19 Stages (at selected moments in time t) of the constant initial functions exponential distribution simulation on the subinterval R^+ presented on the space of instantaneous values $N(t) \in R$ ((**a**): $t = 0$, i.e. initial distribution (**c**): $t = 15$ (**e**): $t = 50$) and on the space $N(t) \times N(t - h)$ in the same moments in time ((**b**): $t = 0$ (**d**): $t = 15$ (**f**): $t = 50$)

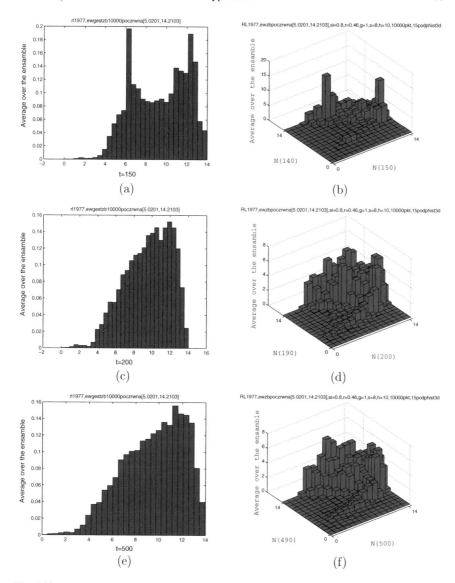

Fig. 6.20 Continued from the Fig. 6.19, in the left column, from top to bottom, subsequent stages of evolution on the space of instantaneous values $N(t) \in R$ (Fig. (**a**): $t = 150$ (**c**): $t = 200$ (**e**): $t = 500$), in the right column, also from top to bottom, on the space $N(t) \times N(t - h)$ (Fig. (**b**): $t = 150$ (**d**): $t = 200$ (**f**): $t = 500$)

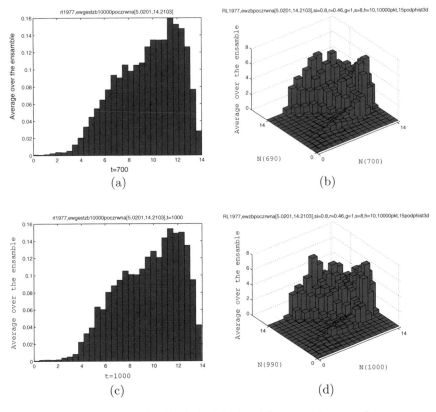

Fig. 6.21 Continued from the Fig. 6.20, in the left column, from top to bottom, subsequent stages of evolution on the space of instantaneous values $N(t) \in R$ (Fig. (**a**): $t = 700$ (**c**): $t = 1000$), in the right column, also from top to bottom, on the space $N(t) \times N(t - h)$ (Fig. (**b**): $t = 700$ (**d**): $t = 1000$

6.3.4 Mixing Properties and Turbulences

The conducted numerical experiments suggest that the flow of the Eq. (6.5) displays typical properties for mixing systems (see Sect. 4.2.2). In particular, all trajectories are unstable (i.e. the system is sensitive to a small change in the initial condition) and the correlation functions for a large set of trajectories and for single trajectories and their time shifts quickly decrease to zero (see Chap. 4.2.2). In addition, the nature of the correlation decay for single trajectories seems to meet the conditions of turbulent trajectories in the sense of Bass definition (see Chap. 4.2.2). Moreover, when we choose adequately the space to visualize the evolution of a large set of trajectories, then its behavior, under the action of the Eq. (6.5), resemble the process of a set "mixing" to some invariant volume. Such behavior is very similar to typical laboratory experiments presenting the mixing process (presenting physical properties of mixing systems) described in physical literature (see, e.g., Dorfman 2001; Lebowitz

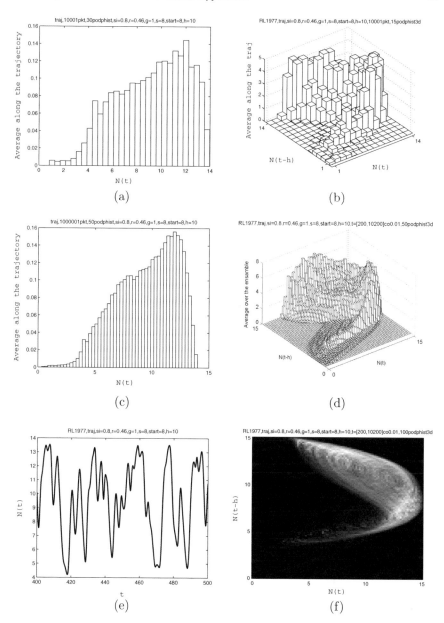

Fig. 6.22 Approximated averages along single trajectory. For 10^4 trajectory points (**a**): on $N(t)$ (**b**): on $N(t) \times N(t-h)$. For 10^6 trajectory points (**c**): on $N(t)$ (**d**): on $N(t) \times N(t-h)$ (**e**): a fragment of the trajectory time course (**f**): histogram on $N(t) \times N(t-h)$ presented from "above" view and in grey scale

Fig. 6.23 Single trajectory of the system (6.5) in the space $N(t) \times N(t - h/2) \times N(t - h)$ fills a similar volume as the whole set of points in this space, on which the system (6.5) operates (compare the Fig. 6.29)

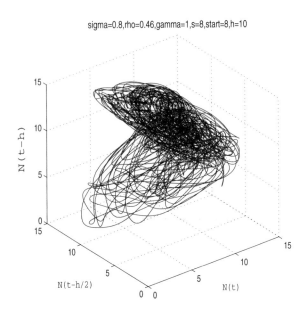

and Penrose 1973). The difference is that in a typical mixing process, points being distributed over the entire available volume (space) reach a homogeneous distribution (see Chap. 4.2.2), here instead the limiting distribution within the invariant volume is not homogeneous. Its numerical approximation in selected subspaces is shown in Figs. 6.21 and 6.22.

6.3.4.1   Trajectory Instability

Figure 6.26 shows the distance changes between two system (6.5) trajectories starting from initial functions with constant values that differ by 0.0001. Changes in the distance between the instantaneous values of the trajectory (Fig. 6.26a) and the distance in the supremum norm (Fig. 6.26b), L^1 (Fig. 6.26c), and L^2 (Fig. 6.26d) were studied, for the Eq. (6.5) solutions considered as a "shifting" state (see Chap. 6.3.2). One can see that at the beginning distances in the norms are very small, then they increase rapidly and continue to oscillate irregularly, exactly as for a typical mixing system (4.13), which we examined in Sect. 4.2.2 (see Fig. 4.10). The graph of the distance between instantaneous values also indicates instability of the trajectory.

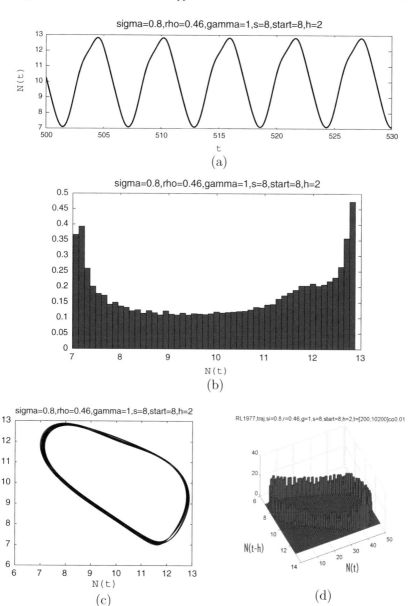

Fig. 6.24 Illustrations of the Eq. (6.5) periodic solution for $\rho = 0.46$, $\sigma = 0.8$, $s = 8$, $\gamma = 1$, and $h = 2$. **a**: Fragment of the trajectory time course. **b**: Histogram of the average along the trajectory on the space of instantaneous values $N(t)$. **c**: Projection of the trajectory on the plane $N(t) \times N(t-h)$. **d**: Histogram of the average along the trajectory on $N(t) \times N(t-h)$

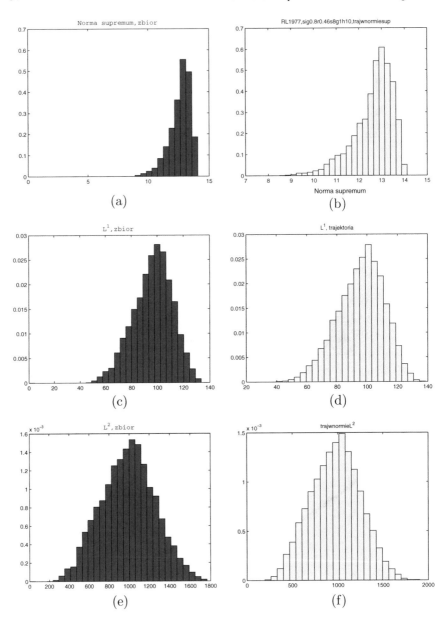

Fig. 6.25 Numerically determined histograms approximating limiting densities for sets of trajectories calculated as a sequence of norms from evolving state (**a**): supremum norm (**c**): L^1 norm (**e**): L^2 norm. Histograms for single trajectories (**b**): supremum norm (**d**): L^1 norm (f): L^2 norm

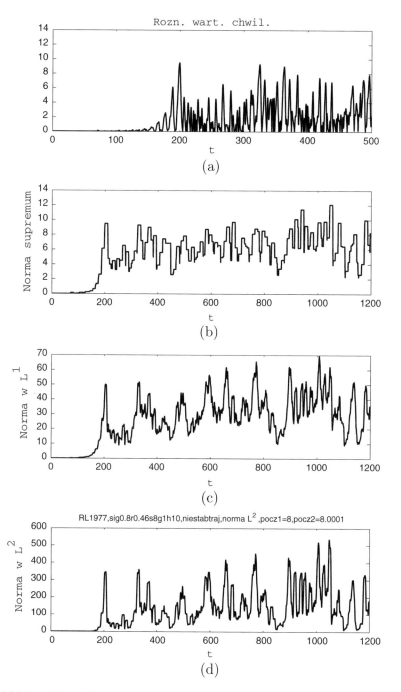

Fig. 6.26 Instability of the trajectory (**a**) in the space of instantaneous values $N(t)$ (**b**) in the supremum norm (**c**) in the L^1 norm (**d**) in the L^2 norm

6.3.4.2 Rapid Correlation Decay and Turbulences in the Sense of Bass

The nature of the correlation function for the trajectory flow and for single trajectories and their time shifts was studied numerically adapting adequately the methodology, presented on the low-dimensional system in the Sect. 4.2.2, to the specifics of Eq. (6.5). Particularly, correlations were studied between receding in time sequences of instantaneous values, and sequences of supremum, L^1 and L^2 norms, for large sets of trajectories and for single trajectories and their time shifts. In each case, the nature of the correlation function resembled typical for mixing systems, i.e. it was quickly vanishing to zero with the increasing time distance τ (see Figs. 6.27 and 6.28).

The lack of correlation indicates also that the attractor has no simple structure e.g., it is not a point or a set attracting periodic trajectories.[1] Let us also notice that the rapid decay of correlation to zero for single trajectories and their time shifts corresponds to the Definition 4.7 of turbulent trajectories in the sense of Bass (see Sect. 4.2.2). We have previously stated that there exist averages along the trajectory (see Figs. 6.22 and 6.25b, d, f), what corresponds to Condition 1 of Definition 4.7. In turn, the decay of correlation to zero with increasing value of time shift τ corresponds to conditions 2 and 3 from the Definition 4.7.

6.3.4.3 Visualization of Mixing Properties

When the space is appropriately selected, we can see how the large set of points is "mixed" within it under the action of the Eq. (6.5). Let us return to the previously considered space $N(t) \times N(t - h/2) \times N(t - h)$. In Fig. 6.29 (discussed above but from a different perspective), we can see selected stages of mixing of a set of 10^4 points with an exponential distribution on a straight line (see Fig. 6.29a). When watching the evolution of this set in the form of animation, it looks exactly as if the initial distributed set was literally mixed, e.g., using a kitchen spoon. Laboratory experiments conducted in order to demonstrate the mixing process, consisting more or less in the fact that a one medium immersed in another medium is evenly distributed over it, are discussed in the physical literature (see e.g., Dorfman 2001; Lebowitz and Penrose 1973). We have a similar situation here, but the distribution of points as a result of mixing is not uniform. Its numerical approximation in selected subspaces is shown in Figs. 6.21 and 6.22.

Formally speaking, we could formulate the conjecture, based on the presented calculation results, that there exist an invariant mixing measure, which is "supported" by an attractor with a non-simple structure for the Eq. (6.5). Probably also almost all trajectories of the system are turbulent in the sense of Bass. Thus, the obtained

---

[1] This conclusion was suggested to me by Prof. Ryszard Rudnicki.

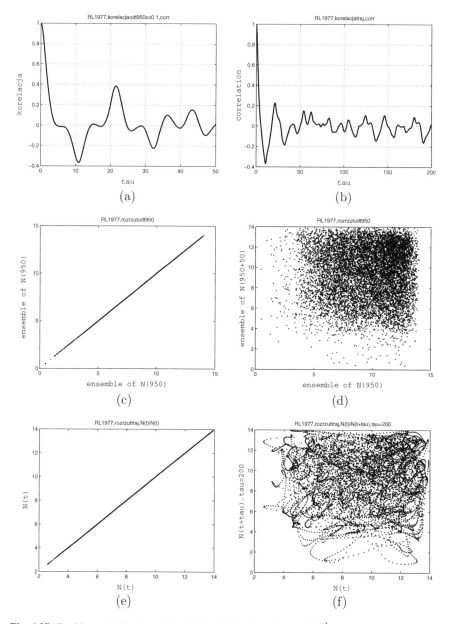

Fig. 6.27 Rapid correlation decay for the Eq. (6.5) **a**: for the set of 10^4 trajectories **b**: for single trajectory and its time shift. **c** and **d**: Spread for set of trajectories and time shifts respectively $\tau = 0$ and $\tau = 50$. **e** and **f**: spread for single trajectory and its time shifts, respectively by $\tau = 0$ and $\tau = 200$

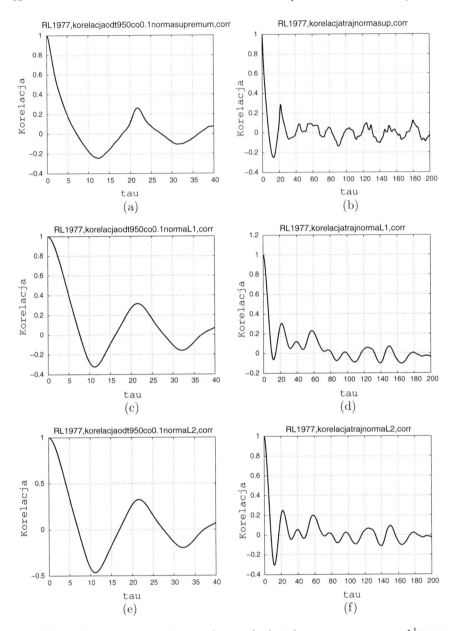

Fig. 6.28 Rapid correlation decay in norms for sets of trajectories **a**: supremum norm **c**: L^1 norm
e: L^2 norm, and for the trajectory and its time shift **b**: supremum norm **d**: L^1 norm **f**: L^2 norm

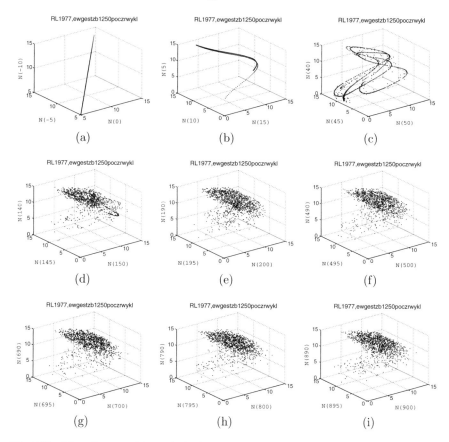

Fig. 6.29 Selected stages of the evolution in the space $N(t) \times N(t - h/2) \times N(t - h)$ of a large set of points under the action of the Eq. (6.5). Points initially distributed exponentially on a line (a) are being mixed up within some volume, which is not changing for longer simulation time

numerical results suggest that the Lasota hypothesis, about the non-trivial ergodic properties of the Eq. (6.5), is true. Additionally, based on the presented analysis, one could suspect that the system is chaotic in the sense of Auslander and Yorke (see sect. 4.2.2).

References

Ackleh, A.S., Deng, K., Ito, K., Thibodeaux, J.: A structured erythropoiesis model with nonlinear cell maturation velocity and hormone decay rate. Math. Biosci. **204**, 21–48 (2006)

Adimy, M., Bernard, S., Clairambault, J., Crauste, F., Génieys, S., Pujo-Menjouet, L.: Modélisation de la dynamique de l'hématopoïèse normale et pathologique. Hématologie Revue vol. 14, no. 5 (2008)

Adimy, M., Crauste, F.: Mathematical model of hematopoiesis dynamics with growth factor-dependent apoptosis and proliferation regulations. Math. Comput. Model. **49**, 2128–2137 (2009)

Adimy, M., Crauste, F.: Global stability of a partial differential equation with distributed delay due to cellular replication. Nonlinear Anal. **54**, 1469–1491 (2003)

Adimy, M., Crauste, F., Ruan, S.: A mathematical model study of the hematopiesis process with applications to chronic myelogenous leukemia. SIAM J. Appl. Math. **65**(4), 1328–1352 (2005a)

Adimy, M., Crauste, F., Ruan, S.: Stability and Hopf bifurcation in a mathematical model of pluripotential stem cell dynamics. Nonlinear Anal. **6**, 651–670 (2005b)

Adimy, M., Crauste, F., Pujo-Menjouet, L.: On the stability of a nonlinear maturity structured model of cellular proliferation. Discret. Contin. Dyn. Syst. **12**(3) (2005c)

Adimy, M., Crauste, F., Halanay, A., Meamtu, M., Opris, D.: Stability of limit cycles in a pluripotent stem cell dynamics model. Chaos Solitons Fract. **27**, 1091–1107 (2006a)

Adimy, M., Crauste, F., Ruan, S.: Modelling hematopoiesis mediated by growth factors with applications to periodic hematological diseases. Bull. Math. Biol. (2006b)

Alvarez, E., Castillo, S., Pinto, M.: (ω, c)-Pseudo periodic functions, first order Cauchy problem and Lasota-Wazewska model with ergodic and unbounded oscillating production of red cells. Boundary Value Prob. (2019). Article number: 106

Alzabut, J.O.: Almost periodic solutions for an impulsive delay nicholson's blowflies model. J. Comput. Appl. Math. **234**, 233–239 (2008)

Anosov, D.V.: Ergodic properties of geodesic flows on closed Riemanian manifolds of negative curvature. Sov. Math. Dokl. **4**, 1153–1156 (1963)

Arnold, V.I.: Mathematical Methods of Classical Mechanics, 2nd edn. Springer, Berlin (1989). (Tłumaczenie z rosyjskiego)

Auslander, J., Yorke, J.A.: Interval maps, factors of maps and chaos. Tôhoku Math. J. II. Ser. **32**, 177–188 (1980)

Bass, J.: Stationary functions and their applications to the theory of turbulence. J. Math. Anal. Appl. **47**, 354–399 (1974)

© The Editor(s) (if applicable) and The Author(s), under exclusive license to Springer Nature Switzerland AG 2021
P. J. Mitkowski, *Mathematical Structures of Ergodicity and Chaos in Population Dynamics*, Studies in Systems, Decision and Control 312, https://doi.org/10.1007/978-3-030-57678-3

Bernard, S., Pujo-Menjouet, L., Mackey, M.C.: Analysis of cell Kinetics Using a Cell Divisin Marker: Mathematical Modeling of Experimental Data. Bipohys. J. **84**, 3414–3424 (2003)

Birkhoff, G.D.: Proof of a recurrence theorem for strongly transitive systems. Proc. Natl. Acad. Sci. USA **17**, 650–655 (1931a)

Birkhoff, G.D.: Proof of the Ergodic theorem. Proc. Natl. Acad. Sci. USA **17**(1931), 656–660 (1931b)

Birkhoff, G.D., Koopman, B.O.: Recent contributions to the ergodic theory. Math.: Proc. Natl. Acad. Sci. **18**, 279–282 (1932)

Bodnar, M., Foryś, U.: A model of immune system with time-dependent immune reactivity. Nonlinear Anal.: Theory, Methods Appl. **70**(2), 1049–1058 (2009)

Bradul, N., Shaikhet, L.: Stability of the positive point of equilibrium of nicholson's blowflies equation with stochastic perturbations: numerical analysis. Hindawi Publ. Corp. Discret. Dyn. Nat. Soc. **2007** (2007). Article ID 92959, 25 stron

Bronsztejn, I.N., Siemiendiajew, K.A., Musiol, G., Muhlig, H.: Nowoczesne Kompendium Matematyki. PWN Warszawa (2004)

Buedo-Fernández, S., Liz, E.: On the stability properties of a delay differential neoclassical model of economic growth. Electron. J. Qual. Theory Differ. Equ. No. **43**, 1–14 (2018)

Carr, J.H., Rodak, B.F.: Atlas Hematologii Klinicznej, wyd. 3. Elsevier Urban & Partner. Original title: Clinical Hematology Atlas, 3 edn., Saunders, an imprint of Elsevier Inc (2009)

Cherniha, R., Davydovych, V.: A mathematical model for the coronavirus COVID-19 outbreak (2020). arXiv:2004.01487v2 [physics.soc-ph] 9 Apr 2020

Chow, S.-N.: Existence of periodic solutions of autonomous functional differential equations. J. Differ. Equ. **15**, 350–378 (1974)

Collet, P.: Short ergodic theory refresher. W Proceedings of the NATO Advanced Study Instytute in International Summer School on Chaotic Dynamics and Transport in Classical and Quantum Systems Cargès, Corsica. Accessed 18–30 Aug 2003. Kluwer Academic Publishers (2005)

Collet, P., Eckamnn, J.-P.: Iterated Maps on the Interval as Dynamical Systems. Birkhäuser, Boston, Basel, Berlin (1980)

Colijn, C., Mackey, M.C.: A mathematical model of hemtopoiesis-I. Periodic chronic myelogenous leukemia. J. Theor. Biol. **237**, 117–132 (2005)

Colijn, C., Mackey, M.C.: A mathematical model of hemtopoiesis-II. Cyclical neutropenia. J. Theor. Biol. **237**, 133–146 (2005)

Craig, J.I.O., McClelland, D.B.L, Ludlam, C.A.: Choroby krwi. W *Choroby wewnętrzne. Davidson.* Tom 3, Redakcja: N.A. Boon, N.R. Cooledge, B.R. Walker, Redakcja edycji międzynarodowej: J.A.A. Hunter, redakcja wydania I polskiego: F. Kokot, L. Hyla-Klekot, Elsevier Urban & Partner (2006)

Crauste, F.: Global asymptotic stability and Hopf bifurcation for a blood cell production model. Math. Biosci. Eng. **3**(2), 325–346 (2006)

Dawidowicz, A.L.: On invariant measures supported on the compact sets II. Universitatis Iagellonicae Acta Mathematica **XXIX**, 25–28 (1992)

Dawidowicz, A.L.: A method of construction of an invariant measure. Ann. Polonici Math. LVII.3, 205–208 (1992)

Dawidowicz, A.L.: Metoda Aveza i jej uogolnienia. Matematyka Stosowana **8**, 46–55 (2007)

Dawidowicz, A.L., Haribash, N., Poskrobko, A.: On the invariant measure for the quasi-linear Lasota equation. Math. Methods Appl. Sci. **30**, 779–787 (2007)

Demina, I., Crauste, F., Gandrillon, O., Volpert, V.: A multi-scale model of erythropoiesis. J. Biol. Dyn. **4**(1), 59–70 (2010)

Ding, J., Li, T.Y., Zhou, A.: Finite approximations of Markov opertors. J. Comput. Appl. Math. **147**, 137–152 (2002)

Ding, J., Du, Q., Li, T.Y.: High order approximation of the Frobenius-Perron operator. Appl. Math. Comput. **53**, 151–171 (1993)

Ding, J., Zhou, A.: Finite approximations of Frobeniu-perron operators. A solution of Ulam's conjecture to multi-dimensional transformations. Phys. D **92**, 61–68 (1996)

Ding, J., Zhou, A.: The projection method for a class of Frobenius-Perron operators. Appl. Math. Lett. **12**, 71–74 (1999)

Devaney, R.L.: An Introduction to chaotic Dynamical Systems. Addison-Wesley Publishing Company, Inc (1987)

Dorfman, J.R.: Wprowadzenie do teorii chaosu w nierównowagowej mechanice statystycznej. PWN, Warszawa (2001)

Von Foerster H.: Some remarks on changing populations. In: Stohlman, F. Jr., The Kinetics of Cellular Plorifelation. Grune & Stratton, New York and London (1959)

Foias, C.: Statistical study of Navier Stokes equations. II, Rendiconti del Seminario Matematico della Universita di Padova **49**, 9–123 (1973)

Foley, C., Mackey, M.C.: Dynamic hematological disease: a review. J. Math. Biol. (2008)

Fomin, S.W., Kornfeld, I.P., Sinaj, J.G.: Teoria Ergodyczna. PWN Warszawa (1987)

Foryś, U.: Matematyka w Biologii. WNT, Warszawa (2005)

Galias, Z.: Metody arytmetyki przedziałowej w badaniach układów nieliniowych. Wydawnictwa AGH, Kraków (2003)

Galias, Z., Zgliczyński, P.: Computer assisted proof of chaos in the Lorenz equations. Phys. D **115**, 165–188 (1998)

Gurney, W.S.C., Blythe, S.P., Nisbet, R.M.: Nicholson's blowflies revisited. Nature **287**, 17–21 (1980)

Górnicki, J.: Podstawy nieliniowej teorii ergodycznej. Wiadomosci Matematyczne XXXVII 5–16 (2001)

Hale, J., Verduyn Lunel, S.M.: Introduction to Functional Differential Equations. Springer, New York Inc (1993)

Hirsh, M.W., Smale, S., Devaney, R.L.: Differential Equations, Dynamical Systems, and an Introduction to Chaos. Elsevier (USA) (2004)

Huang, Ch., Yang, Z., Yi, T., Zou, X.: On the basins of attraction for a class of delay differential equations with non-monotone bistable nonlinearities. J. Differ. Equ. **256**, 2101–2114 (2014)

Kaplan, J.L., Yorke, J.A.: On the nonlinear differential delay equation $x'(t) = -f(x(t), x(t-1))$. J. Differ. Equ. **23**, 293–314 (1977)

Keener, J., Sneyd, J.: Mathematical Physiology. Springer, New York Inc (1998)

Khan, M.A., Atangana, A.: Modeling the dynamics of novel coronavirus (2019-nCov) with fractional derivative. Alexandria Eng. J. In Press (2020). https://doi.org/10.1016/j.aej.2020.02.033

Kucharski, A., Russel, T.W., Diamond, Ch., Liu, Y., Edmunds, J., Funk, S., Eggo, R.M., Sun, F., Jit, M., Munday, J.D., Davies, N., Gimma, A., van Zandvoort, K., Gibbs, H., Hellewell, J., Jarvis, ChI., Clifford, S., Quilty, B.J., Bosse, N.I., Abbott, S., Klepac, P., Flache, S.: Early dynamics of transmission and control of COVID-19: a mathematical modelling study. Lancet Infect Dis. **20**(5), P553–558 (2020)

Kudrewicz, J.: Analiza funkcjonalna dla automatyków i elektroników. PWN Warszawa (1976)

Kudrewicz, J.: Dynamika pętli fazowej, WNT Warszawa (1991)

Kudrewicz, J.: Fraktale i chaos, WNT Warszawa (1993, 2007)

Landau, L.D., Lifszyc, J.M.: Mechanika. PWN Warszawa (2007)

Lani-Wayda, B.: Erratic solution of simple delay equations. Trans. Am. Math. Soc. **351**(3), 901–945 (1999)

de Larminat, P., Thomas, Y.: Automatyka-układy liniowe. t.1 Sygnały i układy, WNT, Warszawa (1983). Original title: Automatique des systemes lineaires.1. Signaux et systems, Flammarion Sciences, Paris (1975)

Lasota, A.: Ergodic problems in biology, Société Mathématique de France. Astérisque **50**, 239–250 (1977)

Lasota, A.: On mappings isomorphic to r-adic transformations. Ann. Polonici Math. **XXXV**.3 (1978)

Lasota, A.: Invariant measures and a linear model of turbulence. Rendiconti del Seminario Matematico della Universita di Padova tome **61**, 39–48 (1979)

Lasota, A.: Stable and chaotic solutions of a first order partial differential equation. Nonlinear Anal. Theory. Methods Appl. **5**(11), 1181–1193 (1981)

Lasota, A., Klimek, A.: Matematyka, czyli opis świata. Interview with Prof. Andrzej Lasota conducted by Andrzej Klimek (2005)

Lasota, A., Mackey, M.C.: Chaos, Fractals, and Noise Stochastic Aspects of Dynamics. Springer, New York Inc (1994)

Lasota, A., Myjak, J.: On a dimension of measures. Bull. Polish Acad. Sci. Math. Math. **50**(2), 221–235 (2002)

Lasota, A., Szarek, T.: Dimension of measures invariant with respect to the Wazewska partial differential equation. J. Differ. Equ. **196**, 448–465 (2002)

Lasota, A., Yorke, J.A.: On the existence of invariant measures for piecewise monotonic transformations. Trans. Am. Math. Soc. **186**, 481–488 (1973)

Lasota, A., Yorke, J.A.: On the existence of invariant measures for transformations with strictly turbulent trajectories. Bull. Acad. Polon. Sci. Ser. Sci. Math. Ast. Phys. **25**, 233–238 (1977)

Lasota, A., Mackey, M.C., Ważewska-Czyżewska, M.: Minimazing theraupetically induced anemia. J. Math. Biol. **13**, 149–158 (1981)

Lebowitz, J.L., Penrose, O.: Modern ergodic theory. Phys. Today (1973)

Lessard, J.-F., Mireles James, J.D.: A rigorus implicity C^1 Chebyshev integrator for delay equations (2019)

Li, T.Y.: Finite approximation for the frobenius-Perron operator, a solution to Ulam's conjecture. J. Approx. Theory **17**, 177–186 (1976)

Li, W.-T., Ruan, S., Wang, Z.-C.: On the diffusive Nicholson's blowflies equation with nonlocal delay. J. Nonlinear Sci. **17**, 505–525 (2007)

Liz, E.: A global picture of the gamma-ricker map: a flexible discrete-time model with factors of positive and negative density dependance. Bull. Math. Biol. **80**, 417–434 (2018)

Liz, E., Röst, G.: On the global attractor of delay differential equations with unimodal feedback. Discret. Contin. Dyn. Syst. **24**(4), 1215–1224 (2009)

Liz, E., Ruiz-Herrera, A.: Delayed population model with allee effect. Math. Biosci. Eng. **12**(1), 83–97 (2015)

Losson, J., Mackey, M.C., Taylor, R., Tyran-Kamińska, M.: Density Evolution Ubder Delayed Dynamics: An Open Problem. Springer, Fields Institute Monographs (2020)

Luo, X., Feng, S., Yang, J., Peng, X.-L., Cao, X., Zhang, J., Yao, M., Zhu, H., Li, M. Y., Wang, H., Jin, Z.: Analysis of potential risk of COVID-19 infections in China based on pairwise epidemic model (2020). www.preprints.org, https://doi.org/10.20944/preprints202002.0398.v1

Mackey, M.C.: Unified hypothesis for the origin of aplastic anemia and periodic hematopoiesis. Blood **51**(5(May)), 941–956 (1978)

Mackey, M.C., Glass, L.: Oscillations and chaos in physiological control systems. Sci. New Ser. **197**(4300), 287–289 (1977)

Mackey, M.C., Milton, J.G.: Feedback, Delays and the Origin of Blood Cell Dynamics, IMA Preprint Series 613 (1990)

Mackey, M.C.: Mathematical models of hematopoietic cell replication and control, strony 149-178 w. In: Othmer, H.G., Adler, F.R., Lewis, M.A., Dallon, J.C. (eds.) The Art of Mathematical Modelling: Case Studies in Ecology, Physiology and Biofluids. Prentice Hall (1997)

McKendrick, A.G.: Applications of mathematics to medical problems. Proc. Edin. Math. Soc. **14**, 98–130 (1926)

Matsumoto, A., Szidarovszky, F.: Asymptotic behavior of delay differential neoclassical growth model. Sustainability **5**, 440–455 (2013). https://doi.org/10.3390/su5020440, ISSN 2071-1050

Mitkowski, P.J.: Uwagi o Równaniu Lasoty-Ważewskiej. Automatyka **12**, 2, 339–347 (2008). Errata (do vol. 12, 2, 2008, p.339) w Automatyka vol. 13, 1, 2009, p. 102

Mitkowski, P.J.: Analysis of Periodic Solutions in Lasota-Ważewska Equation, Conference materials of International Symposium on Nonlinear Theory and its Applications (NOLTA 2008), pp. 57–60, 7–8 September. Budapest, Hungary (2008)

Mitkowski, P.J., 2008. Simulations of Lasota-Ważewska equation. Materiały XIV Krajowej Konferencji Zastosowań Matematykiw Biologii i Medycynie, strony 92–97, Leszno k. Warszawy,

17–20 września, : Editorial Boar: Marek Bodnar. University of Warsaw, Institute of Applied mathematics and Mechanics, Urszula Foryś (2008)

Mitkowski, P.J.: Mathematical Models of Biological Systems. Materiały III Konferencji naukowo-technicznej doktorantów i młodych naukowców. Młodzi naukowcy wobec wyzwań współczesnej techniki, s. 423–427. Politechnika Warszawska 22-24 września (2008)

Mitkowski, P.J.: Numerical analysis of existence of invariant and ergodic measure in the model of dynamics of red blood cell's production system. In: Materiały konferencji: IV European Conference on Computational Mechanics, Palais de Congress, Paris, France. 16–21 May (2010)

Mitkowski, P.J.: Sprzężenia zwrotne w systemach dynamicznych, Nr 82. PAUza Akademicka, Tygodnik Polskiej Akademii Umiejętności w Krakowie, 20 maja (2010)

Mitkowski, P.J.: Chaos w Ujęciu Teorii Ergodycznej w Modelu Zaburzonej Erytropoezy. Doctoral dissertation, Akademia Górniczo-Hutnicza w Krakowie-University of Science and Technology, Cracow, Poland (2011)

Mitkowski, P.J.: Chaos w Ujęciu Teorii Ergodycznej w Modelu Zaburzonej Erytropoezy, Wydawnictwa AGH (AGH University of Science and Technology Press, Cracow) (2012)

Mitkowski, P.J., Mitkowski, W.: Ergodic theory approach to chaos: remarks and computational aspects. Int. J. Appl. Math. Comput. Sci. **22**(2), 259–267 (2012). https://doi.org/10.2478/v10006-012-0019-4

Mitkowski, P.J., Mitkowski, W.: Analysis of the blood production structured model with delay feedback. Materials of CMPD3 2010 The Third Conference on Computational and Mathematical Population Dynamics, p. 174, 31 May–4 June 2010, Bordeaux, France (2010)

Mitkowski, P.J., Ogorzałek, M.J.: Ergodic properties of the model of dynamics of blood-forming system. In: 3rd International Conference on Dynamics, Vibration and Control, Shanghai-Hangzhou (2010)

Mitkowski, P.J., Ogorzałek, M.J.: Evolution of density of states for delay blood cell production model. Conference materials of International Symposium on Nonlinear Theory and its Applications (NOLTA 2010) 5–8 September, Cracow, pp. 71–74 (2010)

Mitkowski, W.: Równania macierzowe i ich zastosowania. AGH University of Science and Technology Press, Cracow (2006)

Mohseni-Moghadam, M., Panahi, M.: Markov finite approximation of Frobenius-Perron Operator for higher-dimensional transformations with a special action matrix. Sci. Iranica **7**(1), 55–56 (2000)

Murawski, R.: Współczesna filozofia matematyki. Wybór tekstów, PWN Warszawa (2002)

Murawski, R.: Filozofia matematyki. Antolgia tekstów klasycznych. Antologia tekstów klasycznych, Wydawnictwa Naukowe UAM Poznań (2002)

Murray, J.D.: Wprowadzenie do biomatematyki, PWN, Warszawa (2006). Tytuł oryginału: Mathematical Biology. I: An Introduction by J. D. Murray. Springer, New York, Inc. (2002)

Musielak, J.: Wstęp do analizy funkcjonalnej. PWN Warszawa (1976)

Myjak, J.: Funkcje Rzeczywiste. Miara. Całka Lebesgue'a, Uczelniane Wydawnictwa Naukowo-Dydaktyczne AGH, Kraków (2006)

Myjak, J.: Andrzej Lasota's selected results. Opuscula Math. **28**(4) (2008)

Myjak, J., Rudnicki, R.: Stability versus chaos for a partial differential equation. Chaos Solitons Fract. **14**, 607–612 (2002)

Nadzieja, T.: Indywidualne Twierdzenie Ergodyczne z topologicznego punktu widzenia. Wiadomości Matematyczne XXXII 27–36 (1996)

Nicholson, A.J.: An outline of the dynamic of animal population. Aust. J. Zool. **2**, 9–65 (1954)

Ogorzałek, M.J.: Chaos and Complexity in Nonlinear Electronic Circuits. World Scientific Publishing, Singapore (1997)

Ogorzałek, M.J.: Design and Implementation of Chaos Control Systems. In: Chen, G. (ed.) W Controlling Chaos and Bifurcations in Engineering Systems. CRC Press, LLC (2000)

Ott, E.: 1997. Chaos in Dynamical Systems, Cambridge University Press, Chaos w Układach Dynamicznych.WNT Warszawa. Dane oryginału (1993)

Parry, W.: Topics in ergodic theory. Cambridge University Press, Cambridge (1981)

Panahi, M.: Approximation measures invariant under piecewise convex transformations. Int. J. Pure Appl. Math. **34**(2), 145–150 (2007)

Pelczar, A., Szarski, J.: Wstęp do teorii równań różniczkowych. Część I. Wstęp do teorii równań różniczkowych zwyczajnych i równań różniczkowych cząstkowych pierwszego rzędu

Peng, L., Yang, W., Zhang, D., Zhuge, Ch., Hong, L.: Epidemic analysis of COVID-19 in China by dynamical modeling (2020). Accessed 16 Feb 2020. arXiv:2002.06563v1 [q-bio.PE]

Penrose, R.: Nowy umysł cesarza. O komputerach, umyśle i prawach fizyki. PWN Warszawa (1996). Original title: The Emperor's New Mind. Concerning Computers, Minds, and the Laws of Physics. Oxford University Press (1989)

Penrose, R.: Droga do rzeczywistości. Wyczerpujący przewodnik po prawach rządzących Wszechświatem. Prószyński i S-ka. Tytuł oryginału angielskiego: The Road to Reality. A complete Guide to the Laws of the Universe (2010)

Prem, K., Liu, Y., Russel, T.W., Kucharski, A.J., Eggo, R.M., Davies, N., Flesche, S., Clifford, S., Pearson, C.A.B., Munday, J.D., Abbott, S., Gibbs, H., Rosello, A., Quilty, B.J., Jombart, T., Sun, F., Diamond, Ch., Gimma, A., van Zandvoort, K., Funk, S., Jarvis, ChI., Edmunds, W.J., Bosse, N.I., Hellewell, J., Jit, M., Klepac, P.: The effect of control strategies to reduce social mixing on outcomes of the COVID-19 epidemic in Wuhan, China: a modelling study. Lancet Public Health **5**(5), E261–E270 (2020)

Prodi, G.: Teoremi Ergodici per le Equazioni della Idrodinamica. C.I.M.E, Roma (1960)

Robinson, C.: Dynamical Systems, Stability, Symbolic Dynamics, and Chaos. CRC Press Inc (1995)

Röst, G., Wu, J.: Domain-decomposition method for the global dynamics of delay differential equations with unimodal feedback. Proc. R. Soc. A **436**, 2655–2669 (2007)

Rudnicki, R.: Invariant measures for the flow of a first order partial differential equation. Ergodic Theory Dyn. Syst. **5**, 437–443 (1985a)

Rudnicki, R.: Ergodic properties of hyperbolic systems of partial differential equations. Bull. Polish Acad. Sci. Math. **33**(11–12), 595–599 (1985b)

Rudnicki, R.: Strong ergodic properties of a first-order partial differential equation. J. Math. Anal. Appl. **133**, 14–26 (1988)

Rudnicki, R.: Wykłady z analizy matematycznej. PWN Warszawa (2001)

Rudnicki, R.: Chaos for some infinite-dimensional dynamical systems. Math. Methods Appl. Sci. **27**, 723–738 (2004)

Rudnicki, R.: O modelu Lasoty-Ważewskiej i jego konsekwencjach. Materiały XXXVI Konferencji Zastosowań Matematyki, Zakopane-Kościelisko, 11–18 września (2007)

Rudnicki, R.: Chaoticity of the blood cell production system. Chaos 19 (2009)

Rudnicki, R.: Modele i metody biologii matematycznej. Część, I: Modele deterministyczne, Instytut Matematyczny Polskiej Akademii Nauk, Księgozbiór Matematyczny tom 2 (2014)

Rudnicki, R., Wieczorek, R.: Świat w równaniach. Academia, Magazyn Polskiej Akademii Nauk Nr **1**(17) (2009)

Sharpe, F.R., Lotka, A.J.: A problem in age-distributions. Phil. Mag. **21**, 435–438 (1911)

Shield, P.: The Ergodic Theory of Discrete Sample Paths. Graduate Studies in Mathematics, vol. 13. AMS (1996)

Silva, C.E.: Wykład wygłoszony na Wiosennej Szkole Układów Dynamicznych w Będlewie, 30 kwietnia - 3 maja (2010)

Sinai, Y.G.: Introduction to Ergodic Theory, Princeton University Press. Princeton. Tłumaczenie z rosyjskiego, New Jersey (1976)

Sinai, Y.G.: Probability Theory. An Introductory Course. Springer, Tłumaczenie z rosyjskiego (1992)

So, J.W.-H.: Dirichlet problem for the diffusive Nicholson's Blowflies equation. J. Differ. Equ. **150**, 317–348 (1998)

Solow, R.M.: The contribution to the theory of economic growth. Q. J. Econ. **70**(1), 65–94 (1956)

Steinhaus, H.: Między duchem a materią pośredniczy matematyka. PWN (2000). Warszawa-Wrocław

Stöcker, H.: Nowoczesne Kompendium Fizyki. PWN, Tłumaczenie z języka niemieckiego (2010)

Swan, T.: Economic Growth and Capital Accumulation. Econ. Rec. **32**, 334–361 (1956)

Szczelina, R., Zgliczyński, P.: Algorithm for Rogorus integration of delay differential equations and the computer-assisted proof of periodic orbits in the mackey-glass equation. Found Comput. Math. **18**, 1299–1332. https://doi.org/10.1007/s10208-017-9369-5

Szlenk, W.: Wstep do teorii gładkich układów dynamicznych. PWN Warszawa (1982)

Tatarkiewicz, W.: Historia Filozofii. Tomy 1-3. PWN Warszawa (1978)

Taylor, S.R.: Probabilistic Properties of Delay Differential Equations. University of Waterloo, Canada, Doctoral dissertation (2004)

Thieme, H.R.: Mathematics in Population Biology. Princeton Series in Theoretical and Computational Biology. Princeton University Press, Princeton (2003)

Tucker, W.: The Lorenz attractor exists. C.R. Acad. Sci. Paris Ser. I **328**, 1197–1202 (1999)

Ulam, S.M.: A Collection of Mathematical Problems. Interscience Publishers Inc, New York, Interscience Publishers Ltd, London (1960)

Ulam, S.M., von Neumann, J.: On combination of stochastic and deterministic processes. Bull. Am. Math. Soc. **53**, 1120 (1947)

Walther, H.O.: (1981) Homoclinic solution and chaos in $\dot{x}(t) = f(x(t-1))$. Nonlinear Anal.: Theory, Methods Appl. **5**(7), 775–788. strony

Walther, H.O.: (1999) The impact on the mathematics of the paper Oscillation and chaos in physiological control systems by Mackey and Glass in Science (1977). arXiv:2001.09010v1 *[math.DS] 24 Jan 2020*

Wang, C., Wei, J.: Bifurcation analysis on a discrete model of Nicholson's blowflies. J. Differ. Equ. Appl. **14**(7), 737–746 (2008)

Ważewska-Czyżewska, M.: Erythrokinetics. Radioisotopic methods of investigation and mathematical approach. Foreign Scientific Publications Department of the National Center for Scientific, Technical and Economic Information, Warsaw, Poland (1983)

Ważewska-Czyżewska, M., Lasota, A.: Matematyczne problemy dynamiki układu krwinek czerwonych. Matematyka Stosowana **6**, 23–40 (1976)

Yi, T., Zou, X.: Global attractivity of the diffusive Nicholson blowflies equation with Neumann boundary condition: a non-monotone case. J. Differ. Equ. **245**, 3376–3388 (2008)

Zhang, J., Peng, Y.: Travelling waves of the diffusive nicholson's blowflies equation with strong generic delay kernel and non-local effect. Nonlinear Anal. **68**, 1263–1270 (2008)

Printed in the United States
by Baker & Taylor Publisher Services